The Cambridge Manuals of Science and
Literature

THE NATURAL HISTORY OF COAL

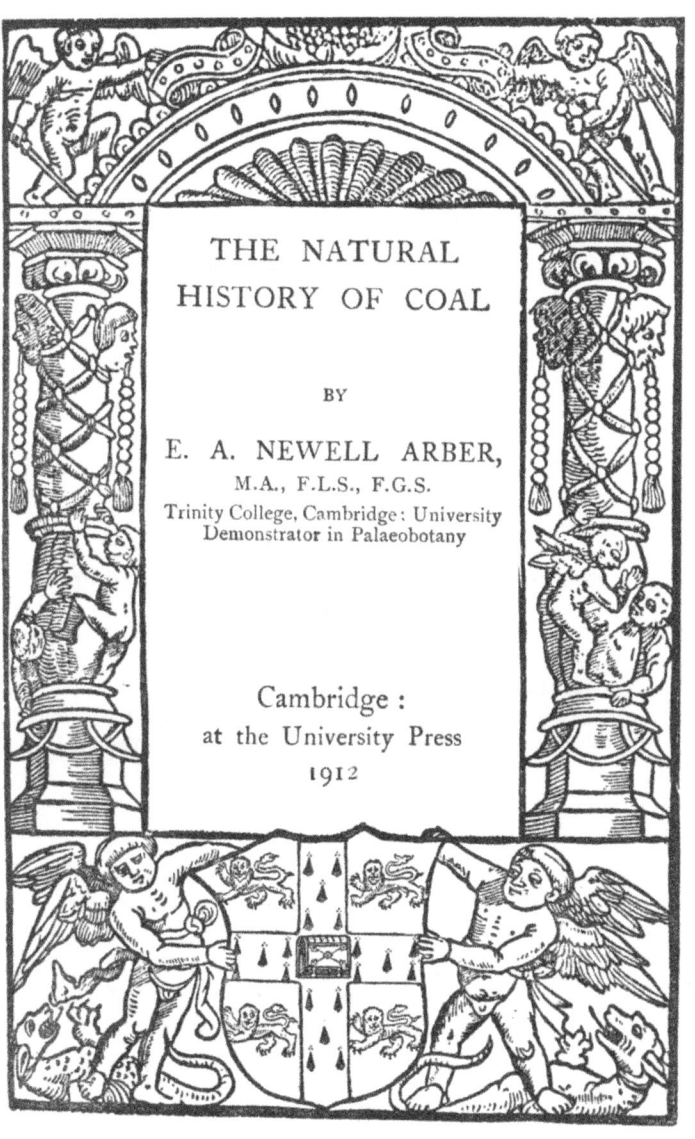

THE NATURAL
HISTORY OF COAL

BY

E. A. NEWELL ARBER,

M.A., F.L.S., F.G.S.

Trinity College, Cambridge: University
Demonstrator in Palaeobotany

Cambridge:
at the University Press
1912

CAMBRIDGE UNIVERSITY PRESS
Cambridge, New York, Melbourne, Madrid, Cape Town,
Singapore, São Paulo, Delhi, Tokyo, Mexico City

Cambridge University Press
The Edinburgh Building, Cambridge CB2 8RU, UK

Published in the United States of America by Cambridge University Press, New York

www.cambridge.org
Information on this title: www.cambridge.org/9781107605763

© Cambridge University Press 1910

First published 1910
Reprinted 1912
First paperback edition 2011

A catalogue record for this publication is available from the British Library

ISBN 978-1-107-60576-3 Paperback

*With the exception of the coat of arms at
the foot, the design on the title page is a
reproduction of one used by the earliest known
Cambridge printer, John Siberch, 1521*

PREFACE

THE substance which we commonly term coal is economically, no doubt, the most important product of the past that we have yet learnt to use. When, however, we turn to the questions, what was the origin of this material, how was it formed, and to other problems of a like nature, it must be confessed that our present knowledge of the applications of coal far outweighs what we have learnt of the past history of this great heritage of long ago. Whether we consider the Geology or the Palaeobotany of coal, we find ourselves confronted with a plexus of problems, either as yet unsolved, or not satisfactorily elucidated at the present time. In this essay, some attempt will be made to enumerate the data which bear on these inquiries, the facts which any theory or theories must explain before they are likely to meet with wide acceptance. The chief difficulties centre around such problems as the original substance from which coal was derived, the conditions under which it accumulated, and the actual mode of conversion of the mother substance into coal.

In the past it would seem that the root of our difficulties in regard to agreement on such points has been the initial hypothesis, that all coals, or at least all the coals of the Carboniferous period, were derived from the same type of mother substance, accumulated under the same conditions, and converted by the same means. It is quite clear that such an assumption is entirely false. There was no one original mother substance, no one original mode of accumulation, and, if all the substances popularly termed coals, whatever their geological age may be, are taken into consideration, no one mode of conversion. Modern research, as we shall see in the chapters which follow, has shown that each case, each seam of coal, must be examined, studied, and judged entirely on its own merits. When this fact has become more widely recognised, it would seem likely that our knowledge of the natural history of coal will progress more rapidly than in the past.

E. A. NEWELL ARBER.

SEDGWICK MUSEUM OF GEOLOGY,
CAMBRIDGE.

November, 1910.

CONTENTS

LIST OF ILLUSTRATIONS

PAGE

THE CHIEF DIVISIONS OF THE STRATIFIED ROCKS.

Tertiary	Recent
	Pliocene
	Miocene
	Oligocene
	Eocene
Mesozoic	Upper Cretaceous Lower Cretaceous
	Upper Jurassic Middle Jurassic Lower Jurassic
	Rhaetic Trias
Palaeozoic	Thuringian ⎫ Saxonian ⎬ Permian Autunian ⎭
	Stephanian ⎫ Upper Carboniferous Westphalian ⎰ (Coal Measures) Lower Carboniferous
	Devonian
	Silurian
	Ordovician
	Cambrian
	Pre-cambrian

(N.B. The lengths of the periods, as shown here, are purely diagram-matic, and only those subdivisions referred to in the text are indicated.)

CHAPTER I

INTRODUCTION. THE CHEMICAL AND PHYSICAL PROPERTIES OF COAL

A KNOWLEDGE of the use of substances, which are popularly termed coals, is no doubt exceedingly ancient. Perhaps the earliest reference, which is supposed to relate to the use of coal as a fuel, is to be found in the book on 'Stones,' by the Greek naturalist and philosopher Theophrastus (about 315 B.C.), where mention is made of an earthy substance, which would kindle, and which could be used by smiths.

Coals are natural products abounding in many parts of the world, and where the beds crop out on the scarped hillsides, the fact that they are not infrequently spontaneously inflammable, may have now and again suggested their use as a fuel, even in regions as yet uncivilised. Nevertheless the use of coal has, for the most part, been confined to civilised or semi-civilised races. As civilisation has matured, so the demand for this substance has increased. The savage of the tropics has little use for coal, even if he knew how to win it. In temperate uncivilised regions, wood in the past has always been a more easily obtained fuel. It is only where the land has become largely deforested, and the march of civilisation has demanded a large supply of compact fuel of high calorific power, that the need for coal has arisen.

The word coal, formerly written cole (A.S. *col*, Icel. *kol*), is not very ancient. It originally implied any substance which could be made use of as a fuel, and was properly added to another word as a suffix. The oldest form is probably 'charcole,' and the word 'cole' for a long period meant simply 'charcoal.' Coal, in the sense in which we now use the term, was originally spoken of as 'pit cole,' 'earth cole,' or 'sea cole,' the last variant being an allusion to the fact that Newcastle coal was from early times brought by sea to London, as it often is to-day.

As wood became scarce, charcoal fell into disuse, and pit coal gradually became more and more important, despite the fact that at various periods Acts of Parliament were passed prohibiting the winning and use of this substance (19). Thus when at length our forests became depleted, the word 'coal' emerged without a rival with which it might be confused.

The existence of coal in various parts of Britain was known from quite early times. The earliest reference to the fact is that found in a Saxon Chronicle of the date 853 A.D. The Newcastle coalfield is perhaps the most ancient coal-winning area in this country, and received its first charter or licence from Henry III in 1239.

In any consideration of the geological problems connected with coal, it is important to remember that this word 'coal' had long been in use prior to the birth of the modern science of Geology. It is a purely popular term, which has crept into scientific terminology. Other words, not wholly dissimilar in significance, have also been used in the past. The term 'Culm,' or in Germany 'Kulm,' meaning the natural slack of an anthracitic coal, has been in common use in the West of England and in South Wales until a comparatively recent period. It, too, has been incorporated into the science, with, as it happens, somewhat unfortunate results.

Coal then is a purely popular term. It includes a vast variety of substances, unlike in geological age, and dissimilar in their chemical and physical properties.

Coal is often spoken of as one of the most important items of our *mineral* wealth. The word 'mineral' is used in a variety of senses, but whatever may be the significance of the above statement from an economic standpoint, scientifically and especially geologically, coal is *not a mineral but a rock*. Coal is as much a rock as sandstones, limestones, granites or marbles. It is further a sedimentary rock, in some cases almost unaltered, while in others, as we shall see, certain substances included in this term are comparable to metamorphosed sedimentary rocks, such as marbles. The mineral substance which constitutes the rock is amorphous, *i.e.* without crystalline form, and consists of hydrocarbons.

We shall defer to the next chapter a consideration of coal as a sedimentary rock. We will, for the present, confine our attention to the physical and chemical properties of coals, evidence which naturally bears on the question of their origin.

There is no such thing as a standard coal. We have to deal with an almost endless series of varieties, some of which pass one into the other. The best known 'coals,' using the term in a popular sense, are Anthracite, the Humic or so-called Bituminous coals, the Sapropelic coals, which include the Cannels and Bogheads, the Brown coals, Lignites, and Peat. Each of these types of fuel are represented by numerous varieties. The Humic coals for instance are particularly varied. Some are excellent for household purposes (house coals), others for the manufacture of coke (coking coals), or gas (gas coals), or again others for raising steam (steam coals).

Steam coals of the Humic denomination merge into smokeless steam coals and Anthracite, and Anthracite into Graphite. Were Graphite capable of being used as a fuel, we should have to include it also in any consideration of the origin of coal.

Lignites are equally varied. In some cases they are loose rather than compact rocks, the structure and form of the wood of which they consist being evident. From such rocks we pass to Brown coals, in which the structure is indistinct, or has been almost completely lost, and the coal is dense and more compact. The substance known as Peat may be a loose, friable mass, showing abundant evidence of its vegetable origin, or it may be compact and structureless, or even pass into Lignites.

These substances may be of very different geological ages, each with a past history dissimilar to that of the other varieties. This is a point to which we shall draw special attention in the next chapter.

The attempt has often been made to limit the application of the word coal to those fuels which are of Palaeozoic age. It would not seem possible however to adopt this suggestion. The whole of these varied substances belong to one great family of rocks, which have at least one character in common, if not more, namely, their vegetable origin. Certainly in any consideration of the natural history of coal, such as the present, it would be extremely unwise, even if it were possible, to confine the discussion to the origin of one particular type of fuel, such as Anthracite or Cannel coal. As we shall see, it is only by a broad survey of the chief problems, which each variety of coal presents, that we can hope to avoid some of the pitfalls into which a hasty and narrow examination of the field might precipitate us.

We have pointed out that there is no standard coal, and that the very word itself is a popular term, which has entered into science. It is therefore impossible to frame any strict, scientific definition of what we mean by a coal. The best diagnosis is perhaps :—a solid, stratified rock, composed mainly of hydrocarbons, and capable of being used as a fuel to supply heat or light or both. That such a definition is extremely imperfect is obvious when we reflect that it does not exclude the oil and bituminous shales, which are not as a rule regarded as coals.

The Chemical Properties of Coal.

The chief constituent of all coals, whatever their geological age may be, is Carbon, which is always greatly in excess of all the other elements present. The presence of Hydrogen is also characteristic. The compounds of these two elements, the Hydrocarbons, form the *combustible matter*. When coal is burnt in excess of oxygen, the carbon forms carbon monoxide and dioxide; the hydrogen partly unites with the oxygen to form water, and partly combines with the nitrogen present to produce ammonia. Nitrogen and oxygen, as well as various other elements, are also present in coal. On burning, these are partly driven off as the *volatile matter*, or remain behind in the *ash*. In addition coal always contains a certain amount of water, the percentage depending on a great variety of circumstances.

The purest forms of coal are those in which elements, other than carbon, hydrogen, oxygen and nitrogen, are few, and present only in small proportions. One of the commonest impurities is sulphur, which is usually present in the volatile products and in the ash, and is derived chiefly from Iron Pyrites (Bi-sulphide of Iron) in the coal. Some coals containing sulphur are unsuitable for certain manufacturing processes of a metallurgical nature. It is also an undesirable impurity even in a house coal, for the occurrence of pyrites in the coal causes the small explosions in the grate, which most people have occasionally noticed, the result being that the coals are scattered beyond the hearth with attendant risk of fire.

Other elements, such as phosphorus and arsenic, occur in some coals, and, when a high percentage is present, render them useless for certain commercial purposes. Other impurities are silica, and silicates, especially those of potash and soda. Various gases are also occluded in the pores of coal. Methane (Marsh Gas) for instance is abundant in Anthracites.

The chief varieties of coal above mentioned (p. 4) differ from one another in their chemical composition, as well as in their physical properties. The average composition of these leading types of coal is shown in the table on p. 9, which is taken partly from Toula's (41) and partly from Walcot Gibson's (12) works. The series begins with an analysis of Beech wood, which offers an interesting comparison with each of the various types of coal. We notice that as we pass from the newer to the older coals, in a geological sense, there is a progressive increase in the carbon percentage, and a corresponding decrease in the percentage of hydrogen, oxygen, and nitrogen. This is a point of importance, on which great stress has been laid in the past, and to which we shall return at a later stage.

It is customary to regard Anthracite as the most perfect stage of coal, because it has a very high carbon percentage. This is a quite misleading conception, and well illustrates the false conclusions to which the contemplation of such a table may give rise. There is very little doubt that Graphite is the ultimate stage of carbonisation. The percentage here is practically 100, for Graphite is pure carbon, and as such is incapable of being used as a fuel.

Substance	Geological age	Carbon percentage	Hydrogen percentage	Oxygen and Nitrogen percentage	Ash percentage
Dried Beech Wood	} Recent	49·89	6·07	44·04	—
Forest Peat		51·47	5·96	32·68	9·67
Moor Peat		53·59	6·33	27·84	12·24
Lignite (Germany) . . .	} Tertiary	57·28	6·03	36·16	·59
Brown Coal (Germany) . .		61·20	5·17	21·28	12·35
Lias Coal (Fünfkirchen, Hungary)	Mesozoic	78·08	3·91	7·32	10·69
Sapropelic Coal (Wigan Cannel)	} Upper Carboniferous	80·07	5·53	10·20	2·70
Humic (or Bituminous) Coal (Newcastle House Coal)		83·47	6·68	9·59	·20
Anthracite (South Wales) . .		91·44	3·36	2·79	1·52

Even from an economic standpoint, the Humic coals, rather than Anthracites, may be regarded as the most perfect fuels, for the high carbon percentage of the latter renders them unsuitable for a vast variety of purposes to which the former are well adapted.

We are thus led to consider the classification of coals on the basis of their chemical composition. This is of great importance from a commercial point of view, and has also a scientific significance. It is not however the only foundation on which a true scientific classification should be based, as we shall see.

From a chemical standpoint, the proportion of carbon to hydrogen, or the $\frac{C}{H}$ ratio as it is called, is important.

The ratios for the substances indicated in the table on p. 9, are as follows

	$\frac{C}{H}$
Dried Beech Wood .	8·22
Forest Peat . . .	8·64
Moor Peat . . .	8·47
Lignite . . .	9·50
Brown Coal . .	11·84
Lias Coal . .	19·97
Sapropelic Coal . .	14·48
Humic Coal . .	12·48
Anthracite . . .	27·21

The varieties of these main types of fuel may be also classified chemically. Those of the Humic and Anthracitic coals have been grouped into genera and species, the former being determined by the hydrogen, the latter by the carbon percentage. (See Seyler (36), Strahan and Pollard (40).)

The chemical basis of classification is no doubt the best scientific index to the grouping of coals, which is available at present. In the future we may hope to combine with it other considerations based on a more perfect knowledge of the origin of the various types of coal; but the time for this has not yet arrived. We may however as a temporary expedient, which possibly in the end will prove to be inaccurate, base our main classification on the supposed *mode of deposition* of the varied substances, which we are here including under the term coal. The following scheme, which will be adopted here, will no doubt meet with strong opposition at the hands of those who hold that all coals originated in much the same way, and it must therefore be regarded simply as the personal belief or opinion of the author.

Terrestrial
> Peat
> Lignites (in part)
> Brown Coals (in part).

Lacustrine (Fresh water)

> Sapropelic coals (Cannel coal, Boghead, Tor-
> banite)
> Humic (the so-called Bituminous) coals (in part)
> ? Anthracite.

Estuarine (Brackish or salt water)

> Lignites (in part)
> Brown coals (in part)
> Humic coals (in part)
> ? Anthracite.

Before leaving the subject of the classification of
coals, we must not neglect the commercial method of
grouping the varieties of Sapropelic and Humic coals,
which is based partly on their chemical, and partly
on their physical properties. Coals yielding large
percentages of volatile products are useful for certain
purposes, especially gas making, but not for others.
The percentage of ash is a very important factor, a
large amount of ash being undesirable even in an
ordinary house coal. A coal, which under incomplete
combustion yields a large amount of coke, is highly
valued on this account. Again, coals almost free
from sulphur, phosphorus, or arsenic are specially
adapted to certain processes. For steam raising
purposes—perhaps the most general use to which
coals are put—the heating or calorific power is of

chief importance, and this depends on the amount of carbon and hydrogen present.

Other considerations of economic value depend upon physical characters. Specific gravity is a factor to be considered, where coal has to be carried in the confined space of a ship's bunkers, or on the tender of a locomotive. The friability or tendency to produce slack has to be taken into consideration, if the coal is to be transported for long distances.

It is of course often possible to obtain the same quantity of heat from a Brown coal as a Humic coal, but the quantity of the former necessary is huge in comparison with that of the latter. It is wasteful, difficult to handle, takes up much more space in storage, and leaves a large amount of ash, which does not facilitate complete combustion. Naturally these matters are of importance from a commercial standpoint.

The Physical Characters of Coals.

We now turn to a consideration of the physical characters of the chief types of coal, beginning with Peat.

Peat is a compact, fibrous substance, sometimes dark brown in colour, sometimes quite black, in which but few traces are to be seen by superficial inspection of the plants of which it is composed. It is a fuel which varies greatly in its characters. It is accumulating to-day in large areas in Russia, Germany, and France as well as in Great Britain and Ireland. We shall defer until a later stage (pp. 54—62) a more special consideration of its mode of formation, and of the conclusions which have been drawn from the facts relating to the origin of peat, and applied to illustrate the genesis of other types of fuel.

Peat beds vary from a few inches in thickness to 30 feet or more. Peat always contains a large proportion of moisture, and, even when dried, it burns slowly, and is expensive as a fuel on account of the small quantity of heat given off per mass, and the large amount of ash and earthy impurities, which hinders complete combustion.

Brown coals and Lignites. These two terms are used by many authors as synonymous, while others distinguish Lignites from Brown coals, as a distinct type. The latter is the view we shall adopt here, although as is well known the two grade into one another.

A *Lignite* may be defined as a coal in which the dominant constituent consists of wood, so little altered that the form and structure are evident on inspection. It is a fuel of a fibrous character, derived from slightly altered wood, with which is also associated a considerable amount of structureless material, as well as many decomposed, but still recognisable fragments of bark and leaves. There are many varieties, ranging from loose to compact fuels. Compact Lignites are often called Earthy Lignites, owing to the tendency of the wood to crumble to dust.

Lignites are rare in British rocks as deposits of any thickness. The most extensive are those of Bovey Tracey in Devonshire (see p. 37), not far southwest of Exeter. These Lignites consist chiefly of the wood of Coniferous trees, believed to be allied to the Sequoias, now represented by the Redwood and Mammoth trees of California. The beds are many feet in thickness, and are interstratified with clays and sands. They have been known and worked since 1714.

Thin, local accumulations of Lignite occasionally occur in Eocene beds in the south and east of England. Abroad, vast deposits of Lignite occur in the Rocky Mountain region of the United States, and elsewhere in Cretaceous and Tertiary sediments. Such coals are practically unknown from Palaeozoic rocks.

Brown coal is a dark brown or black, compact fuel, which does not show any obvious evidence of woody structure. It more closely resembles a very hard and compact Peat, or a Sapropelic coal, from which however it differs chemically (see p. 9). It is a very variable fuel, some forms, which are however very rare, approximate closely to the Humic coals or even to Anthracite, and are known as Bituminous Brown coals, or Pitch coals. These are black in colour, and usually possess a conchoidal fracture.

Brown coals do not occur in Britain, but they are widely distributed in the Mesozoic and Tertiary rocks of Europe, Asia and North America (see p. 37).

Sapropelic coals. We now pass to consider coals which are almost exclusively of Palaeozoic age. The Sapropelic coals are of two main types, Cannel coals and Bogheads. The latter are also known as Torbanites in Scotland. These fuels differ from the Humic coals, not only in the large proportion of volatile constituents which they contain, which render them excellent for gas manufacture, but in their physical properties.

Cannel coal is a black, dull (not shining) coal, clean to the touch (does not soil the fingers), breaking with a conchoidal fracture. It burns readily with a candle-like flame (hence the name 'candle'- corrupted into 'cannel'-coal) and leaves little ash. In some respects it resembles jet, and will take a high polish, but it never exhibits any trace of woody structure. There are many varieties. Splint coal is a cannel possessing a slaty fracture; Parrot coal, a variety found in one of the Scotch coalfields, which crackles when burnt. Cannel often occurs in seams

of Humic coal, and nearly always in the form of lenticular bands.

Bogheads and Torbanites are rare but very interesting coals, confined to a few localities in Scotland,
so far as Britain is concerned, and occurring also in
the coalfield of Autun in Central France, in Russia
and the United States.   Bogheads are also found in
New South Wales and South Africa, where they are
associated with the *Glossopteris* flora (see p. 34).
This type of coal was first worked at Torbanehill
(Linlithgowshire) in the middle of the last century,
hence the name Torbanite.   It differs so much from
an ordinary coal in appearance, that, in 1853, a famous
lawsuit took place in Edinburgh, with the object of
deciding whether it was really a coal or not.   The
lessees of Torbanehill at that time had power to
work coals, but not minerals, and it was claimed by
the owners that Torbanite was a mineral, and not a
coal.   The courts however decided that Torbanite
was a coal, and that the lessees were within their
rights.

Bogheads closely resemble Cannel coals in appearance.   They are rich in volatile hydrocarbons, but
unlike Cannel coal, the ash percentage is high.

The Humic or so-called *Bituminous coals.*
Ordinary house and steam coals are usually spoken
of as bituminous coals, a most unfortunate term,
since it suggests that they consist largely of the

mineral bitumen, which is not the case. I propose to speak of such coals here as Humic coals, a term already used by David White (42) and others.

The appearance of ordinary Humic coals is familiar to every one in the form of house coal, and needs little description here. They do not present clean surfaces (*i.e.* they soil the fingers), in parts they have a shining lustre, and they break into more or less rectangular lumps, coinciding with the bedding planes and joints.

Such coals however are usually composite. They consist of alternating layers, possessing different physical properties. One layer consists of a dull black substance, of silky or fibrous texture, and has the appearance of having undergone charring. This is called the *mother of coal* or *mineral charcoal* (Fr. *fusain*, Ger. *faserkohle*). In these dull layers, charred fragments of vegetable tissues, such as leaves, bark or wood, may occasionally be clearly perceived. With these dull layers alternate laminae of bright coal with a shining lustre, in which as a rule no trace of any fossils can be seen. The chemical composition of the two layers also differs. Bright coal contains less carbon, and leaves less ash than dull coal.

The different varieties of Humic coals, such as steam coals, coke and house coals, do not differ markedly in their physical properties, though they are unlike in their chemical composition. Some approach Anthracites in their chemical character (semi-anthracites), while others are less perfect coals, not far removed from certain types of Brown coal.

Anthracite, or Stone coal as it is sometimes called, differs entirely in appearance from the Humic coals. It is clean to the touch (*i.e.* does not soil the fingers), possesses a shining, silvery, semi-metallic lustre, and breaks with a conchoidal fracture. It has to be raised to a high temperature, usually by the aid of another type of fuel, before it will burn, and its flame is feeble or only slightly luminous, and practically smokeless. In other words the percentage of volatile matter is very small, and this is also true of the ash. It gives more heat per mass than any other type of coal.

Culm is a word used locally in Devonshire, also in Pembrokeshire and elsewhere in South Wales, for the natural slack of an anthracitic coal. In these regions the rocks have undergone great earth movements, and the result has been that the coal has been largely crushed to powder. In Devonshire, near Bideford, this powder has long been used as the basis of black paint.

CHAPTER II

COAL AS A ROCK, AND THE ASSOCIATED ROCKS

COAL, as we have seen, is, in a geological sense, a rock and not a mineral. There is abundant evidence for the view that coal belongs to the group of sedimentary, as opposed to igneous, rocks. While some coals such as peats are terrestrial accumulations, others correspond more closely to aqueous sediments, such as sandstones, limestones or shales, though of course they are entirely dissimilar in their chemical and physical properties.

Wherever a seam of coal is found, and whatever its geological age may be, the seam is always a minor constituent of the sequence of rocks to which it belongs. In the earlier period of the study of Geology, such deposits were sometimes spoken of as *accidental beds*, and, though this term is really inaccurate, it has the merit of implying that such seams are only of minor geological importance in the series. Even in the Coal Measures (Upper Carboniferous), which are the chief coal-bearing series in

England and Wales, it is the sandstones, shales, and clays which are the dominant lithological types, and not the coals. Among the other minor sediments occurring in this series are calcareous rocks, whether in the form of persistent or impersistent bands or of concretionary nodules, both of which are greatly varied as regards the conditions of their formation. Bands of ironstone or of ironstone nodules also frequently occur. Siliceous deposits of chert or gannister are rarer.

The attempt has not infrequently been made to discuss the origin of a seam of coal without special reference to the associated rocks. This is extremely unwise, for the evidence presented by the sediments interstratified with the coals has undoubtedly a most important bearing on some of the vexed questions directly connected with the origin of coal itself. A seam of coal represents but one of a series of epochs, which must be studied as a whole, if we are to understand exactly what has taken place in the past.

We will here consider the chief features of the rocks associated with the Upper Palaeozoic coals. As regards the more recent coals, which we shall then pass on to discuss, the same conclusions for the most part hold good.

There is no reason to suppose that the Coal Measure sandstones, limestones, and stratified shales or unstratified clays, with the exception of the underclays of the coal seams, differ, as regards their mode of origin, from the precisely similar deposits belonging to other geological formations of later date. Yet their nature and origin is extremely varied, and the same is true of the corresponding sediments of the Mesozoic and Tertiary periods.

Some sandstones are freshwater (fluvial or lacustrine) deposits, while others were laid down under littoral, lagoonal or even marine conditions. The shales and clays are equally varied. Some are of

Fig. 1. The order of deposition of river-borne sediments on the floor of a lake after a recent flood. After Lapworth.

freshwater origin, while others are even more highly marine and deeper water sediments than the associated sandstones, as is shown by the characters of their faunas. The ordinary sequence of deposition, whether freshwater or marine, of littoral sandstones grading into shales or muds is too widely recognised to need detailed explanation here (Fig. 1). It will suffice to add that the dominant rocks of the Carboniferous or of any other geological period offer no exceptions to the ordinary rules of sedimentation, as illustrated by the rivers and coast lines of to-day.

The lithological characters of the rocks themselves are not always to be relied upon as a sure indication of the conditions and circumstances under which they were deposited. For evidence of this nature we must look rather to the included organisms, whether animals or plants. Where the shales or limestones contain certain types of fish or Cephalopod (Goniatite) remains, we may conclude that the rocks were deposited under marine conditions. Where the fauna consists of freshwater genera of fish and Mollusca, we have a clear indication of a non-marine facies. Plant remains are as a rule indicative of non-marine conditions, though not necessarily of freshwater deposition. Many plants, especially of the Carboniferous period, probably flourished chiefly in estuarine or lagoonal habitats. At the same time we must remember that such remains do occasionally occur in undoubtedly marine sediments.

The Sandstones.

The sandstones of the Carboniferous period are extremely varied, alike as to their lithological characters, and their fossil contents. At one end of a series we find coarse grits, in which the component particles of silica are sharp and angular. The Millstone grit of the Midland counties, which is a somewhat exceptional and local development of the lower beds of the Lower Coal Measures, is of this type. Other sandstones present all stages in a graduated series, in which the grain of the rock becomes finer and finer. The finer sediments frequently cease to be purely arenaceous, and, as the admixture of silicates of alumina increases, they may pass eventually into sandy shales, or typical argillaceous deposits.

The sandstones vary in thickness from massive deposits, hundreds of feet thick, to bands of an inch or less in extent, alternating with shales. As viewed in one particular quarry or cutting, they may appear to form regular beds, but, if they are traced laterally, it will often be found that they are really lenticular masses, thinning out to a few inches in thickness at either end, and closely interwoven with associated rocks of other types. On the other hand, some of the more massive series may continue uniformly over an area many miles in extent, and where this is the case they form important landmarks in the coalfield.

In colour the sandstones vary from white, through grey to yellow, and even to a deep red tint. In England there are two main types: the Grey Measures and the Red Measures, so called from the hues of the prevailing sandstones. The Grey Measures occur in almost every coalfield, and are the chief coal-bearing or productive measures. The sandstones of these series are usually grey or yellow. The Red Measures, which are often associated with the Grey

Measures, though always belonging to a higher horizon, rarely contain coals of economic value. These sandstones contain much iron, and are stained a deep red colour. Such Red Measures occur in the Cumberland, South Lancashire and Yorkshire coalfields, and are perhaps best developed in the Midland coalfields of Staffordshire and Warwickshire.

As we have indicated, the Red Measures always occupy a higher horizon than the Grey Measures, with which they are associated. In the occurrence of such beds we have a clear indication of a general change in the conditions, which prevailed during the latter portion of the Upper Carboniferous period, as compared with the earlier represented by the Grey Measures. The red beds are regarded as evidence of the setting in of desert conditions, unfavourable to coal formation, and these conditions apparently continued in force and reached their maximum in Permian and Triassic times in Western Europe. It need scarcely be pointed out that any evidence as to changes of conditions, such as is furnished by the sandstones of the Coal Measures of England, cannot be neglected in an attempt to unravel the mysteries which surround the origin of the different types of coal. The fact that only a very few, and very thin seams of coal occur in the Red Measures indicates that, while the conditions which prevailed at the time when the Grey Measures were deposited were

highly favourable to a rapid accumulation of vegetable material, as the climate became more and
more desert-like, the circumstances were less and less
conducive to vegetation, and consequently to coal
formation. This conclusion is supported by the
fact that, while in the latter portion of Westphalian
times (see p. x), to which the whole of the Upper
Carboniferous series of Britain may be attributed,
we see clear indications of the setting in of desert
conditions in some parts of England, yet this
change was after all merely of a local nature. In
certain districts of France and Germany no such
change took place, and the conditions continued
favourable for the deposition of Grey Measures, not
only throughout the Westphalian and Stephanian
epochs (see p. x), but also in Permian times.

The Shales and Clays.

The term 'shale' is applied to laminated rocks,
which consist chiefly of silicates of alumina. They
are consolidated muds or silts, and split readily along
the bedding planes. The clays are similar but un-
laminated.

The shales of the Coal Measures are, as a whole,
more constant in thickness and composition than the
sandstones, with which they are associated, as we
should expect when we remember the conditions of

deposition. Yet they vary enormously in their physical characters. Pure shales, known to the miner as 'Binds,' pass into sandy shales ('Rock-binds' or 'Stony-binds') and even into impure argillaceous sandstones. Some shales split into splintery flakes, others break with a conchoidal fracture, others again irregularly. In thickness they vary from an inch to many feet. In some districts they are more important than the sandstones, as the dominant rock type, in others the converse holds.

The Underclays or Fireclays.

In the English coalfields, the beds immediately above the coal seams are generally shales, though sandstones are not infrequent in this position. Below the coals, in this country and elsewhere, there usually occur beds of unstratified clay, known to geologists as *underclays*. Fireclays is another term frequently applied to these beds, from the fact that they furnish excellent materials for the manufacture of firebricks. In addition, a host of other names are used locally by miners, such as 'spavin' in Yorkshire, 'seat-earth' or 'warrant' in Lancashire. In France and Belgium however underclays have no special designation.

The occurrence of underclays below coal seams was first emphasised by Logan (26) in 1840, who stated that they always occur below the coals of the South Wales coalfield. This is also the rule elsewhere, though it is not invariably the case. Lyell (27) noticed that underclays occur in the Great

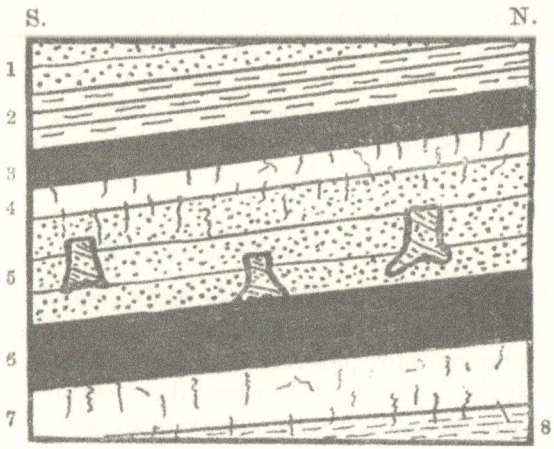

Fig. 2. Diagrammatic section of coal-bearing rocks. 1. Sandstone. 2. Shales forming roof of seam. 3. Seam of coal. 4. Unstratified underclay with *Stigmaria*. 5. Sandstones forming roof of seam, with *Stigmaria* above, and three upright trunks, the internal layers of which do not correspond to the sandstones. One trunk is seated on the top of the seam of coal, 6. 7. Unstratified underclays with *Stigmaria*. 8. Bed of shale penetrated by *Stigmaria*. The beds are dipping south at about 15°.

Appalachian, and other coalfields in the United States. It is also asserted that such deposits occur beneath certain Mesozoic coals, such as those of the Yorkshire Coast and of Fünfkirchen in Hungary (14), the former being of Oolitic, the latter of Liassic age.

On the other hand, fireclays are frequently met with without any coal seam within many feet of them. They may also be found above a coal seam, though this is a rare phenomenon. Coal seams may occur immediately above a conglomeritic sandstone as in South Wales, above limestones as in some of the Scotch coalfields, or above igneous rocks as in Central France.

The characters of these underclays vary greatly. They are sometimes rather soft, unctuous, greyish clays, at other times, fairly hard and sandy. In the Lower Coal Measures of Yorkshire and Lancashire, the underclays of some of the lower seams are highly siliceous, and form a hard, flinty rock. The silica percentage may be as high as $96\,^0/_0$, and is rarely lower than $57\,^0/_0$. These siliceous unstratified rocks are known as Gannister, and the series to which they belong is spoken of as the Gannister beds.

The fireclays of the Coal Measures are characterised by the fact that they are usually, but not always, full of *Stigmaria* (Figs. 12 and 13), the rhizophore of various Palaeozoic Lycopods. As a rule these rhizophores and their roots penetrate the underclays in all directions, and under the hammer the rock, being quite unstratified, breaks irregularly according to the direction of the fossils which it contains. The abundance of *Stigmaria* and Stigmarian roots in underclays has an important bearing on the theories of the formation of coal, which we shall consider in a later chapter.

Concretions and Lenticular Bands.

Concretions and lenticular, impersistent bands of limestone, or of ironstone, abound in the Coal Measure sequence. The simplest case is where nodules of a concretionary nature are formed, either in sandstones or shales, by a local infiltration of calcium carbonate, or carbonate of iron. The nodules are frequently unfossiliferous. At other times they contain an abundant flora or fauna. From the included fossils we learn that these nodules are of very diverse origin. Some of them, containing numerous plant remains (either in the form of impressions, or more rarely as petrifactions) or presenting a varied fauna of freshwater or estuarine Mollusca, were obviously not derived from rocks deposited under marine conditions. Other nodules again contain a typical marine fauna of fish and Goniatites, and but few, or badly preserved, plant remains.

Not infrequently, in place of a series of nodules embedded in sandstones or shales, we find thin, impersistent, lenticular bands of impure limestone, or of ironstone. These are probably to be regarded simply as infiltrations of calcium or ferrous carbonate, on a large scale as compared with the nodules. These limestone bands also vary as to the conditions of formation ; some are marine, others estuarine or freshwater. Beds of ironstone are also found to pass laterally into limestone, and *vice versa.*

While such lenticular beds are far from infrequent in the Upper Carboniferous rocks of Britain, when we turn to the Lower Carboniferous series, we find a great sequence of massive limestones in both England and Scotland. Those occurring in England are almost entirely deep-water, marine formations, and do not contain coal seams. As however we near the Scottish border, we find evidence of a change in the conditions under which these sediments were deposited, and in Scotland coal seams are associated with the Carboniferous limestone, which is there of an estuarine or freshwater facies.

From even such a brief review of the varied lithological types associated with the Carboniferous coals of Britain, it is obvious that a great variety in the conditions of deposition of the associated rocks must have existed, and this must be borne in mind when we come to discuss the origin of coal itself. The same applies equally to the newer coals of Mesozoic and Tertiary age—the Brown coals and Lignites. They are everywhere interstratified with sedimentary rocks of various types, and of diverse origins. Even the modern peats may rest on old, denuded land-surfaces, and in some cases lie buried beneath the Boulder Clays of the Glacial epoch.

The Geological Age of Coals.

There is another fact, which must constantly be kept in mind when considering the origin of coals, especially when the term is used in a broad and unrestricted sense. These substances, and the rocks with which they are associated, are of very different Geological ages. Both Palaeozoic, Mesozoic and Tertiary coals abound, and further, coals are found in a very large number of the subdivisions of these great periods (see table opposite p. 1) as we shall shortly point out. When we reflect how very different has been the geological history of these rocks, we readily perceive the truth of the maxim, that all coals have not been formed in the same way. For instance, the Palaeozoic coal-bearing rocks of England have been subjected to great pressures, as the result of earth movements, which took place towards the close of that period. The Tertiary Lignites of Bovey Tracey in Devonshire, or still later deposits, such as the recent Peat, have not been influenced by such movements or pressures. This fact alone is not without significance, when we attempt to explain and to account for the differences presented by these three types of fuel.

It is sometimes imagined by those who have not looked carefully into the matter, that all coals, using the term in the restricted sense of a house-coal, are of Carboniferous age, perhaps because in England the Carboniferous rocks are practically the only coal-bearing series of economic importance, with the exception of modern Peats. This view is however quite unfounded. If we include Graphite in our consideration, although it is not a fuel, then coal of one sort or another occurs in each main division of the stratigraphical series (see p. x) and in almost every subdivision, in one part of the world or another. Even if we exclude Graphite, coal is found in rocks of every age, which are later than the Silurian. From an economic standpoint, however, the coals of the Palaeozoic formation are the most valuable, and the most widely distributed, especially those of the Upper Carboniferous and Permian periods.

We will now briefly review some of the more important coals belonging to different geological periods or their subdivisions (see table, p. x).

Pre-cambrian to Silurian. If we regard some Graphites, which are practically pure carbon, as the ultimate results of the metamorphism of coals such as Anthracites, then this rock is undoubtedly the oldest member of the family in a geological sense, even though it is no longer capable of being used as a fuel. Enormous deposits of Graphite, some of which are probably of vegetable origin, occur in the Pre-cambrian of Canada, in the Ordovician of the Lake District, in the Devonian of Cornwall, and the Carboniferous of the Alps.

Devonian Coals. The oldest known Humic coals are those from the Upper Devonian of the north-east portion of Bear Island in the Arctic regions, where seams of 3 ft. 6 ins. in thickness occur.

Carboniferous coals. Coals of varied types are abundant in the Carboniferous rocks of Europe, Northern Asia, China and North America. They do not occur uniformly throughout the series in any one region, but vary locally as to the horizon. In India, and the Southern Hemisphere, including Australia, South and South-East Africa and many parts of South America, especially Brazil, Argentina, and Chili, we have a facies of Upper Carboniferous and Permian rocks distinct from that which existed at the same period in the Northern Hemisphere. These regions then formed part of a vast continent, the flora of which was in large part dissimilar to that of Europe and North America, and is distinguished as the *Glossopteris* flora. Coals are associated with *Glos-sopteris*-bearing rocks in all the portions of this southern continent.

In Great Britain coals both of Lower and Upper

Carboniferous age are important. In Scotland, some of the chief coals belong to the Lower Carboniferous, others to the Westphalian division of the Upper Carboniferous. In England and Wales all our coals are of Westphalian age. In France and Germany not only Westphalian but Stephanian coals occur, in the coalfields of Commentry, St Étienne, and the Saar region.

In the United States the most important coals are Westphalian in age.

Permian coals. Coals of Autunian age occur in several of the small coalfields of central France, such as Autun, Blanzy and Brive, and also in the Saar basin in Germany. Coals of Saxonian age occur in Saxony near Dresden and in Bohemia. Coals of Permian age also occur in the Southern Hemisphere as already mentioned.

Triasso-Rhaetic coals. Coals of Triasso-Rhaetic age are found in the small coalfield of Richmond in Virginia (U.S.A.), in the Lettenkohle of the Keuper of Germany, which is however scarcely of economic importance, and in Tonkin and Japan. It is stated that some of the Japanese coals of this age are Anthracitic.

Jurassic coals. We have three examples of coals of Jurassic age in Britain, all of which have been used at various periods. One of these coals is still worked and has been so for nearly 100 years past. Coal beds occur in the Lower Oolite of the Yorkshire coast, in the Lower and Middle Estuarine Series of Robin Hood's Bay and Cloughton Wyke. These coals are only a few inches thick, though occasionally they swell out to a thickness of 6 to 10 inches. Some of them are said to be of very fair quality, and were worked in a small way in the early part of the last century. Underclays are found beneath some of the seams.

Coal of Jurassic age also occurs in the extreme North-East of Scotland, at Brora in Sutherlandshire, and has been worked intermittently, chiefly since 1811. The occurrence of coal in this locality is said to have been known so far back as 1598, and 70,000 tons had been extracted by the year 1827. The coal resembles a Humic coal of Carboniferous age in appearance, but yields a high percentage of ash. The main coal is from 2 to 3 feet thick, and contains much iron pyrites. The roof of this coal is a hard calcareous sandstone, while the floor is a stiff, dark coloured or black clay, sometimes bituminous.

The Kimeridge coal of Kimeridge Bay in Dorset is, by some, regarded as a bituminous shale, though it resembles the Sapropelic coals of Carboniferous age in appearance. The bed is about 2 feet thick, and is interstratified with the Kimeridge clay. It has often taken fire spontaneously. It has been worked at various times, but without much success, as a source of paraffin, and even gas. It is useless as a household fuel on account of its very unpleasant smell.

Other Jurassic coals occur at Fünfkirchen in Hungary (Liassic age) (14), in the neighbourhood of the Caucasus in Russia, in Japan (anthracitic), and possibly in Borneo.

Cretaceous coals. Brown coals and Lignites of Cretaceous age are well developed in the North-Western regions of the United States (Rocky Mountain region) and in Alaska. They also occur in Japan, where some of them belong to the Upper Cretaceous.

Eocene coals. Lignites or Brown coals of this period occur in the United States, and probably in New Zealand and Sumatra.

Oligocene and Miocene coals. Oligocene Lignite occurs at Bovey Tracey in Devonshire, and was once used locally as a fuel. Lignites and Brown coals occur in Oligocene or Miocene deposits in Germany, Hungary, Russia, Japan and elsewhere.

Recent coals. Peat is abundant in Northern Europe, and extends to the polar regions.

CHAPTER III

IT is agreed that coal has been derived as the result of the alteration of some accumulation or deposit, which we will call the *mother substance.* Of what did this mother substance consist? The balance of opinion has always inclined to the view that it was composed of vegetable debris in whole or in part, and this conclusion is almost universally accepted to-day. On the other hand, there have been those who hold that some coals at any rate were of inorganic origin, and others who attribute to animal origin the greater portion of the bulk of certain types of fuel.

If we include in the term coal such terrestrially formed accumulations, as Peats and Lignites, then there is no doubt whatever that some coals are of vegetable origin. But when we turn to the older coals of the Palaeozoic period, we find that the original substance has become so greatly altered that very little trace of its primary structure can be seen. In some cases we find only a homogeneous, amorphous mass, and it is partly for this reason that doubt has been thrown on its vegetable origin.

It has been concluded that some Palaeozoic coals are of an inorganic nature and have resulted from eruptions of the mineral bitumen. It is supposed that in some cases these eruptions were submarine, or sublacustrine, that sheets of bitumen poured out in lava-like spreads on the floors, of estuaries or lakes, and that in this way seams of coal were formed. These sheets were then covered up with ordinary sediments, such as sands and muds, and at a later stage fresh eruptions of bitumen appeared, and in this way further seams were initiated. The presence of plant impressions in the coal itself, and in the roof and floor of the seams, is explained by the hypothesis that the bitumen penetrated and incrusted such fragments of vegetable debris as they lay at the bottom of the estuary or lake, and thus casts or incrustations of the plants in bitumen were formed. The arguments against this view are however overwhelming. If coal is really a bitumen, then it is quite unlike in its chemical and physical properties any other substance known under that name. Further, if coal seams have resulted from bituminous eruptions, we should expect to find veins or perhaps dykes of bitumen ejected into, and penetrating the associated rocks (sandstones, shales and limestones) in all directions. Nothing of the sort occurs however. Also, if plant remains were preserved in bitumen, we should find that the bitumen had penetrated into the cells and intercellular spaces, where the structure has been preserved, and this is not the case. Further, if casts of plant remains were preserved in bitumen, they could hardly have withstood transportation, and we have ample evidence that some such casts, and also coal debris have been transported for some distance.

It is doubtful whether the bitumen hypothesis to-day finds more than a very few supporters. From time to time, however, various workers, vaguely dissatisfied with the prevailing theories, or perhaps overcome by the difficulties which surround the study of the subject, appear to have taken refuge in the inorganic hypothesis of the origin of coal, despite the almost overwhelming evidence, which has accumulated during the last hundred years, in favour of the vegetable genesis of most, if not all, coals. Such a case has occurred fairly recently. A French writer, Rigaud (33), has revived the bitumen theory in a modified form. He assumes that the climate of the Carboniferous period was tropical. This is an assumption, which, as we shall see (p. 69), is at the best 'not proven,' so far as it rests on the evidence of the vegetation. Since, with few exceptions, a tropical climate and peat growth are to-day incompatible, it is therefore suggested that Carboniferous

coals could not have been formed from peat, which
is apparently the first stage in the formation of all
coals according to this author. M. Rigaud thus
falls back on the theory of bituminous eruptions,
and in favour of his contention he states that, on
chemical analysis, some substances such as phos-
phoric acid and certain alkalies, which should be
present, in his opinion, in the ash of coals, if they
are formed from a former growth of vegetation,
are lacking. He endeavours to show that, had
these substances been present originally, they could
not have been removed by the solvent or other
action of water, during the process of coal formation. It may however be pointed out that, even
if the facts are as M. Rigaud asserts, it is quite
impossible to build any broad hypothesis of general
application to the origin of coals, on such minute
and relatively unimportant details, without leaving
out of all account a great mass of evidence to the
contrary, as he appears to have done.

The view that animal matter may have contributed to the bulk of certain coals of the Sapropelic type, which has recently been advocated by
Prof. Potonié (28), and to which we shall return
at a later stage (p. 125), applies only to certain special
cases, and is not antagonistic to the view that coals
generally have been formed from a mother substance
of vegetable origin.

The Vegetable Origin of the Mother Substance.

We now return to the evidence which tends to show that the mother substance of coal was originally of a vegetable nature.

In rare and exceptional cases, in certain coals of Palaeozoic age the original mother substance is so little altered that its nature may still be recognised. One of the best known instances which was first described by Gœppert (13) in 1848, and more recently by Zeiller and by Renault (30) is the Papier-kohle or Toula coal of Russia (Fig. 21). In the mines of Malovka and Tovarkovo, in the province of Toula (Tula) in Central Russia, there is a seam known by the above names, which consists entirely of sheets of the finely striated bark of an important Carboniferous tree, *Bothrodendron.* These sheets of bark, or rather the cuticle of the bark, are very loosely united together, and so easily separated that it is said they may be loosened, and blown apart and away by the wind. Here the original substance appears to have undergone very little alteration.

M. Grand'Eury (15) has also pointed out that some coals in the coalfield of St. Étienne, in Central France, are homogeneous deposits made up of the bark of *Cordaites,* or of species of *Calamites.*

In the case of modern Peats, and Tertiary Lignites

and Brown coals, as we shall see in the following
chapters, it is easy to recognise the vegetable nature
of the deposit.

Next, we have the evidence of plant impressions
and petrifactions, occurring actually in the coal itself.
In this country impressions of plants are on the
whole very uncommon in this position, and no
experienced collector would ever dream of searching
in the coal itself for well-preserved examples of the
flora. They do however occasionally occur, the bark
of *Sigillaria* being perhaps the most frequent fossil,
while examples of fern-like fronds have been also
found. In some of the coalfields of the Continent,
plant impressions in coal are of more frequent
occurrence. Gœppert has described from the coalfields of Upper Silesia seams of coal, which appear
to be largely, if not entirely, composed of Sigillarian
wood, and between every layer impressions of the
bark of this plant, sometimes associated with leaves,
may be recognised. In Bohemia, also, fern-like leaves
have been described as occurring in Humic coals.

Next, we have instances of casts of plants, the
hollow pith cavities of which have become filled
with mud or sand, giving a solid pith cast in sandstone or shale, while the woody tissues have been
converted into coal. Where the pith-casts are characteristic of the genus, as in *Calamites* and *Cordaites*,
there is no difficulty in explaining exactly what has

taken place in such cases, since the anatomical structure of these genera is known. Specimens of this nature are by no means infrequent, and have been repeatedly figured by various authors. Here we have absolute evidence that coal may be derived from the wood of a Calamite or a Cordaitean stem. The great contraction in volume which the coal represents, as compared with the diameter of the original wood, is another interesting point in connection with these specimens. When the pith-cast is stout, or where we know from structure specimens that the particular stem was of considerable size, and the wood well developed, the rind of coal, which now surrounds the pith-cast, is often quite small and narrow.

In the case of the plant petrifactions found in the nodules, known as coal-balls, in certain seams in Yorkshire and Lancashire—a very interesting subject which we shall discuss at length at a later stage (p. 106)—we have undoubtedly cases where part of the original mother substance of the coal seam has been petrified by infiltration of calcium carbonate, while the rest has been converted into coal. Since the calcareous nodules, which have resulted from this infiltration, are entirely composed of fragments of plants, there is every reason to believe that the rest of the seam was of a similar nature.

Thus the evidence that the mother substance

of coals was of a vegetable nature is, quite apart from the results of a microscopic study of the structure of coals, extremely strong.

The Microscopic Nature of Coal.

If we prepare thin slices of coal—a task often far from easy on account of the friable nature of the material—and examine them under the microscope, we shall find as a rule that they are very disappointing as regards the amount of information we can obtain from them. Such sections are usually opaque, even when quite thin, and the substance is obviously very homogeneous. So much so that some thirty years ago a large work was published, in which are given many figures of cracks in the material running in various directions, intersecting one another and filled with a slightly more transparent substance than the main mass of the coal. To these cracks and other purely inorganic irregularities in the homogeneity of the coal, a too vivid imagination ascribed an organic origin of an algal or other nature!

Despite the difficulties of this line of investigation, it has received attention at the hands of many observers from the time of Witham (1833) and Link (1838) (25) onwards. The late Prof. W. C. Williamson of Manchester devoted much time, during the latter portion of his life, to forming a large collection of sections of various coals from all parts of the world, but the results were never published. So far the microscopic examination of coals, unaided by preliminary chemical treatment, has not proved a success.

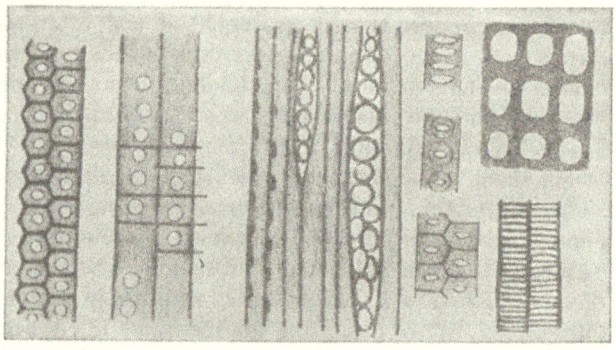

Fig. 3.   Fragments of woody tissue, especially pitted and scalariform tracheides, obtained from macerated, Humic coal.   Magnified. After Dawson.

Better results have been obtained where some solvent, such as ether, has been employed. The chemical preparation adopted by Link (25) was to break the coal into splinters, and then to render them transparent by the use of petroleum.   Or, if the coal be macerated by treatment with sodium carbonate, the method used by Schmid and Schleiden

(1846), or better still by Schultze's method (1855) of immersing coal in a mixture of nitric acid and potassium chlorate, followed by treatment with ammonia, most coals can be made to yield fragments of tissues or of cuticularised structures, such as the epidermis, sporangia, or spores. The latter are practically indestructible, and in recent years the revival of Schultze's method, as applied to fossil impressions, which are only represented by a carbonised film, has led to very important results in the hands of Prof. Nathorst of Stockholm.

In 1859, Dawson (7) showed that the layer of mineral charcoal (p. 18), which usually alternates with the layers of bright coal in the case of Humic coals, if treated chemically, can be made to yield scalariform tracheides and other fragments of woody tissue (Fig 3). In 1870, Huxley, in an article in the *Contemporary Review*, pointed out that the 'Better Bed' coal of Bradford, Yorkshire, was built up chiefly of sporangia and spores. Williamson afterwards showed that Huxley's sporangia were really megaspores, and since then a large number of varieties of both megaspores and microspores have been described from British coals, especially by Kidston, Wethered and others. Newton has also concluded that Tasmanite or Australian white coal, which is of Tertiary age, is largely composed of spores.

The discovery that certain coals, especially Sapropelic coals of the Cannel type, were largely composed of the spores of Lycopods aroused great interest at the time, and may have led to somewhat rash conclusions as to the origin of coals in general,

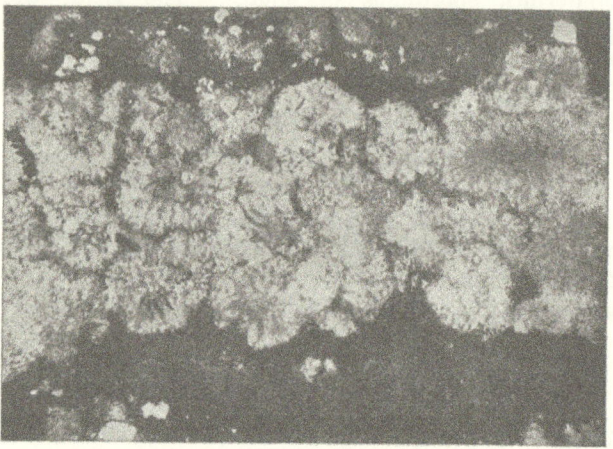

Fig. 4. Section through a Boghead coal, from Autun in France, showing that it consists largely of aggregations of the oval or circular thalli of *Pila*. Considerably enlarged. After Bertrand.

it being vaguely supposed that, because Cannels were built up of spores, other coals were probably of a similar origin. When this idea was put to the test however, it was soon found that no trace of spores could be found in many coals. Prof. Williamson examined specimens of the Lower Oolite coal of Yorkshire, but failed to find any evidence of spores. In more recent years, however, we have learnt that the Sapropelic coals, which include the Cannels and Bogheads, are a race apart from other coals. They differ essentially in the nature of the original mother substance, as the brilliant work of Renault and Bertrand has made clear. The earlier discoveries of the occurrence of Lycopod spores in Cannels have been followed by equally interesting observations of the nature of the Boghead and Torbanites, with which the Cannels are now grouped as a class.

The Scotch Boghead or Torbanite was first scientifically examined by Bennett in 1857, who noticed that it contained numerous orange-yellow or bright red transparent bodies of irregular outline, embedded in a dark opaque matrix. Fischer and Rüst (1882) regarded them as resinous patches, and it was not until 1892 that Bertrand and Renault (4) put forward an entirely new explanation of the origin and nature of the materials of which such Bogheads consist. These authors made a careful microscopic investigation of a Boghead from Autun in Central France, and they concluded that the bright, orange-yellow patches were really the remains of gelatinous algae, to which they gave the name *Pila bibractensis* (Fig. 4). These microscopic thalli

(Fig. 5) are irregularly spherical, or ellipsoidal, and contain from 600—700 cells, pyramidal in shape and

Fig. 5.   Two thalli of *Pila bibractensis* from the Boghead of Autun (France).   Magnified 270 times.   After Bertrand.

radially disposed, especially near the periphery of the thallus. The same genus was afterwards recognised in Bogheads from Scotland and Australia, and in some cases three-fourths of the bulk of the coal consisted of *Pila*, embedded in a structureless brown matrix.

In 1894, the same authors (5) examined the Kerosene shale of the Permo-Carboniferous of New South Wales, which they concluded was built up of a somewhat similar alga, *Reinschia* (Fig. 6). Here the thallus was saclike, composed of a single layer of cells surrounding a central cavity. In some cases nine-

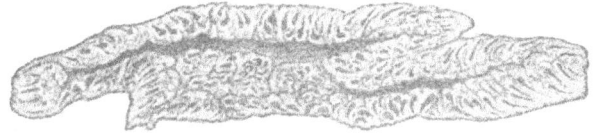

Fig. 6. A thallus of *Reinschia australis* from the Kerosene shale of New South Wales. Magnified 315 times. After Bertrand and Renault.

tenths of the Kerosene shale consists of thalli of *Reinschia australis* (2). Both *Reinschia* and *Pila* are commonly associated with pollen grains of various Carboniferous plants (3).

Despite the immense amount of labour and attention devoted to a study of the structure of the Sapropelic coals at the hands of Prof. Bertrand and the late Bernard Renault, it may perhaps be doubted if the evidence obtained bears out all the conclusions claimed from it. It will be generally

admitted that the care and patience of these observers have shown that Boghead coals are largely
made up of structures, which are probably of
vegetable origin, though we may not be prepared
to agree that the algal nature of *Pila* and *Reinschia*
has been completely demonstrated. It may be so,
but these fossils, whatever their origin may be, are
admittedly greatly altered as compared with their
original state, and this alone warns us that we must
not draw too rigid conclusions from the present
evidence.

As a matter of fact the algal nature of Boghead
coals has recently been disputed by Prof. Jeffrey
(23) of Harvard. He has applied a new chemical
treatment to coals, by making use of a mixture
of nitric and hydrofluoric acids. He finds that the
supposed algae of Renault and Bertrand are really
only the megaspores of Vascular Cryptogams. 'The
imagined algae are in fact only the pores in the
strongly sculptured coats of the spores in question.
The apparent organization of the algae as colonies
forming a hollow sphere, is explained as the highly
sculptured wall of the macrospore surrounding its
empty cavity.'

Thus the whole question of the algal nature of
Pila and *Reinschia* has been reopened, and it will
be very interesting to see whether Prof. Jeffrey's
conclusions are confirmed. If they are, then the

affinity of Bogheads and Torbanites to Cannel coals becomes even closer than was formerly suspected. It has been believed that as a rule the Cannels are built up of spores, with which are associated few or no alga-like thalli, and that intermediate coals may be found, grading to true Bogheads, in which the supposed thalli increase in number, and the macro- and microspores decrease, until at last in true Bogheads the former alone prevail.

From this brief review of a very large subject it is clear that some evidence of vegetable origin has been recognised in all the main types of coal, with the exception of Anthracite, and thus there are good grounds for the belief that the mother substance from which coal has originated consisted of an accumulation or deposit of vegetable origin.

CHAPTER IV

THE MODE OF ACCUMULATION OF TERRESTRIAL COALS

We have now to approach the difficult question of the mode of accumulation of the mother substance from which coal-seams have resulted. As has been already pointed out, we include peat under the term coal. The conditions of formation of this fuel are exceptionally favourable for study, since peat is being formed at the present day on a large scale in Northern Europe. We will therefore commence with a brief review of the manner of its occurrence, in the hope that it may throw light on the origin of other types of coal.

Peat is a terrestrial accumulation of vegetable debris, laid down under water which is not free or circulating. The mode of formation of this fuel has been widely studied in many countries, especially in Sweden and Germany, and an enormous literature exists on the subject. In recent years our Scottish peats have received special attention at the hands of

Mr Lewis (24), to whose work we shall here particularly refer.

Peat is an accumulation which is formed with
extreme slowness, and the extent of the beds varies
from a few inches to 30 feet or more. All observers
agree that a bed of peat is a composite formation,
consisting of layers of very different dominant
plants, accumulated under quite varied conditions.
Thus a section of a peat bog represents a chapter in
the geological history of the district, which in some
cases carries us back as far as the Glacial period.

The upper layers of a peat-moss, as such deposits
are called in Scotland, consist of living plants growing
on a substratum of a former, partly altered vegetation,
which is saturated with water held between the mesh-
work of the debris, like a liquid in a sponge. The
water-loving plants flourishing on the surface vary
in different districts, and they have also varied at
different epochs in the history of the same region, as
we shall see shortly. From among the many varieties
of living peat, we can recognise two main types, the
Upland or Mountain, and the Lowland.

Upland or Mountain peats are found in many
districts in Northern Europe, and also at considerable
elevations in the Alps. In England and Scotland,
they are common on the hills of the Pennine range,
and in the Grampians. In Scotland generally, peat
formation is practically confined to the western half,

which is the region of maximum rainfall, though very little fresh peat is being formed there to-day, as Mr Lewis (24) has shown. The surface layers of

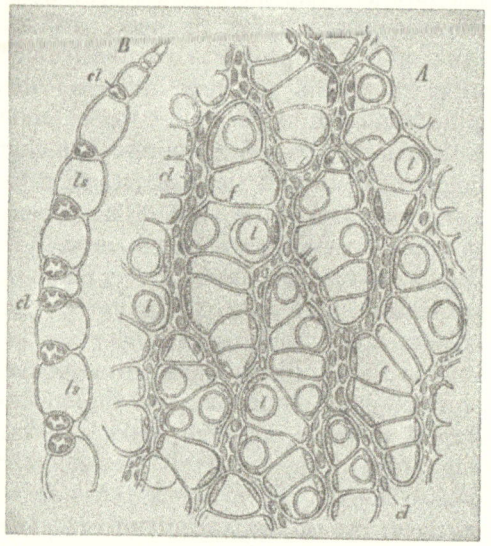

Fig. 7. The leaf of the bog-moss, *Sphagnum*, *A* in surface view, *B* in transverse section. The small tube-like cells, *cl*, contain chlorophyll; the large empty cells, *ls*, have spiral bands, *f*, and holes, *l*. In the living state the large cells contain water. After Gœbel.

upland peats consist of the bog-moss *Sphagnum*, and many other plants flourish on the surface, especially

Heaths (*Erica*). The leaves of *Sphagnum* consist chiefly of large, water-retaining cells (Fig. 7). The dead plants of *Sphagnum* and other surface vegetation sink through the spongy layers of living bog-moss and collect beneath them, thus contributing to the mass of subfossil peat below. When forested areas are invaded by peat-mosses, the decaying trunks of the water-logged trees also sink into the peat and contribute to its bulk (Fig. 18). This is well seen in progress to-day in the large Alder marshes of East Prussia.

Lowland peat is formed in marshy meadows, on the borders of lakes, and on the banks of streams. The dominant plants forming the living surface layers are not bog-mosses, but many different Monocotyledons, members of the Cyperaceæ and Grasses, such as *Cladium* and *Phragmites*. Mosses do not always play an important part, though species of *Hypnum*, but not *Sphagnum*, may frequently occur. (38 Introduction). The peats of the English Fenland or of Western Ireland offer excellent examples of lowland peats, and vast deposits of a similar nature occur in Russia, Sweden, and Germany.

The next point which we have to consider is the nature of the alteration which the vegetable debris undergoes, when it sinks to the bottom of the living surface layers. This change is quite distinct from decay on dry land in the open air. The debris is almost entirely shut off from the atmosphere by the water above it, and modifications of a different nature take place, which result in the formation of hydro-carbons and of humic, ulmic and other organic acids. In a short time an accumulation of a heterogeneous character is converted into a more or less homoge-neous, compact, brown pulp, the constituent elements of which are no longer recognisable, except after special and detailed examination.

These changes are brought about chiefly by the action of bacteria, as Renault has shown. The preliminary mechanical dissociation of the debris is probably aided by the action of fungi, and also by certain animals which live in such habitats, such as worms. Next follows the disappearance of all the thin-walled cells and tissues—the work mainly of the bacteria. Both the walls and the contained proto-plasm are reduced to a brown, structureless pulp, in which the harder and more resistent tissues are embedded. The degree to which this process is carried varies under different circumstances. It may be arrested at a comparatively early stage, or, on the other hand, the walls of the thick-walled elements may be considerably eroded, before the mass becomes unsuited to the existence of bacteria, probably on account of the excess of humic and other acids set free. Thus the final product may be either a more or less amorphous, brown, or black, paste-like mass,

or it may contain obvious traces of the vegetation
from which it is derived.

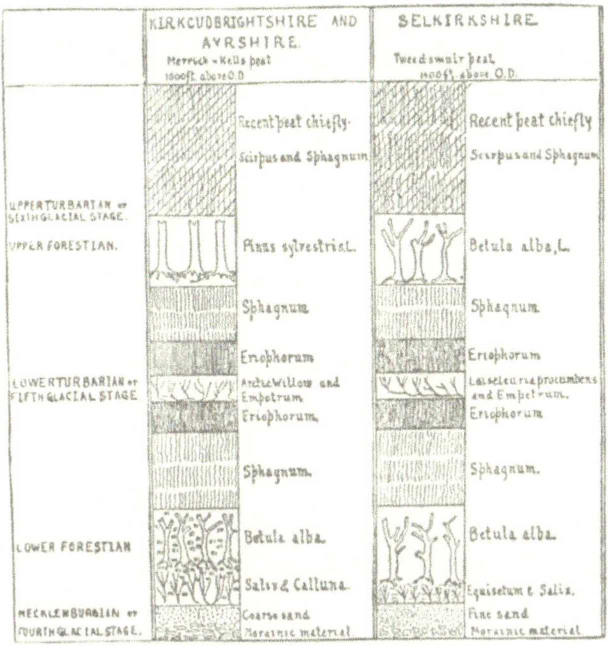

Fig. 8. Diagrammatic sections of Peat-mosses in Scotland, showing
the varied origins of the successive layers.   After Lewis.

The character of the resulting peat varies
according to the nature of the plants, which have

contributed to its substance. This leads us to a study of the origin of the various layers seen in a section of a peat-moss. As we have already indicated each bed of peat has its own special history. Diagrammatic sections of two peat-mosses, which date back to the Glacial period, are shown in Fig. 8, taken from Mr Lewis' memoirs (24). The fact that such accumulations are quite ancient is not surprising, when we remember that peat grows very slowly, the formation of 2 feet in a century being a generous estimate.

Mr Lewis (24) states that the sequence of the peat in the South of Scotland, the Highlands and the Hebrides is as follows :

> 7 Recent Peat
> 6 Upper Forestian
> 5 Upper Peat Bog
> 4 Second Arctic Bed
> 3 Lower Peat Bog
> 2 Lower Forestian
> 1 First Arctic Bed
>   Glacial Deposits.

One of the most striking features is the alternation of forest beds, which are imperfect Lignites, with beds of peat proper. In Scotland there are two distinct forest beds, the Lower Forestian consisting chiefly of trunks, branches and bark of the Birch (*Betula alba*),

and the Upper Forestian similarly composed of
Scots Pine (*Pinus sylvestris*). In Sweden there
are three forest beds composed respectively of Pine,
Oak (*Quercus robur*), and Spruce (*Picea excelsa*).
These forest beds may be so dense, and the timber
so large, tough, and closely packed, that they are

Eriophorum
Calluna
Sphagnum
Salix reticulata
Betula nana

Betula alba

Betula nana

Salix reticulata

Peat formed from aquatic plants
underlying the arctic bed.

Fig. 9. Diagrammatic section of a Peat-moss in the
Shetland Islands. After Lewis.

difficult to cut through, even with the aid of an axe
and saw.

It is obvious that, in a district where the peat
represents such a series of epochs, the conditions
must have varied at different periods. At one time
they favoured the formation of peat, at another the

climate changed, the area became drier and thickly forested. These and other alternations have been repeated more than once. A fact which is perhaps even more strikingly emphasised in such diagrams of peat sections, as are shown in Figs. 8 and 9, is that the dominant vegetation forming not only the forest beds, but also the peat proper, has varied from time to time. In the former case the older forests consisted of Birch, in the latter of Scots Pine. The peat proper has at one time been formed chiefly by *Sphagnum* and *Eriophorum* (Cotton-grass), at others by *Scirpus*, and at others again by dwarf Arctic Willows, such as *Salix reticulata*, and *S. herbacea*, associated with the trailing Azalea (*A. procumbens*) and other Arctic plants.

Ancient Terrestrial Accumulations.

The study of modern peat deposits carries us back to the Glacial period, but no further. We have now to inquire whether we can recognise more ancient deposits of a like nature, formed under similar conditions. In such an inquiry we must remember that similar deposits are also in process of formation at the present day, though under somewhat different circumstances. In the case of peat, we have an aqueous accumulation, formed under conditions of which the most important is standing, and non-circulating water. Swamps, as we shall see later, may furnish almost identical products, but under different conditions, for their waters are free and circulating. It is therefore obvious that, even if we may attribute to some of the Lignites and Brown coals an origin akin to that of the layers of peat and forest beds, exhibited by a section of a peat-bog, other similar accumulations may have originated under swamp conditions. Thus it is an extremely difficult matter to determine whether, in a particular case, the genesis was terrestrial or estuarine.

The first essential condition to the formation of peat is the existence of an abundance of vegetation, capable of flourishing under the particular circumstances of a water-logged soil. There are many areas in various parts of the world where there is no peat, although the conditions would appear to be extremely favourable to the formation of such accumulations. Darwin long ago pointed out that this is the case in many districts in South America, and that the absence of peat there may be attributed to a lack of peat-forming vegetation. Mr Lewis has recently shown that in the west of Scotland, where the rainfall and other conditions appear to be favourable to the growth of peat, little or no peat is being formed at the present time. Thus the occurrence of peat depends not only on the prevalence of particular climatic conditions, but also on the existence of a special type of vegetation in the district.

It must also be remembered in attempting to trace ancient terrestrial accumulations, that any arguments from the new to the old, based on our knowledge of recent peat, can be only justified so long as the types of vegetation concerned have remained the same. We have seen that in the case of peats, a very large number of plants of diverse affinities play important parts. In addition to Mosses, various Angiosperms, both Dicotyledons and Monocotyledons, are important in this respect. The associated forest beds are derived partly from Dicotyledons, and partly from Coniferae. All these classes of plants have existed thoughout the Tertiary period, and also in the upper portion of the Cretaceous. We may therefore expect to find in the rocks of these periods accumulations parallel to those presented by the various layers of a peat-bog. Prior however to the Upper Cretaceous, there were no Angiosperms. Coniferae certainly existed throughout Mesozoic times, and the same may be also true of the Mosses, but it is almost beyond doubt that neither of these groups, and still less the Angiosperms, were in existence in Palaeozoic times. Thus on these grounds alone we are not justified in arguing directly from a modern peat to a Palaeozoic Humic coal. The nature of the vegetation undoubtedly determines the character of the product, and Palaeozoic plants were almost entirely distinct from those which contribute to the bulk of any modern fuel.

So far however as the Tertiary and Upper Cretaceous rocks are concerned, we are justified in our search for more ancient accumulations, of similar origin to modern peats. Some of the Brown coals and Lignites of these periods certainly present a parallelism to our peats and forest-beds. The former resemble hard, compact, dry peats, the latter may be regarded as equivalent to the woods of a forest-bed, which have undergone some further alteration. No doubt both the Brown coals and the Lignites have suffered a greater degree of modification than has been the case with the peats and forest beds, as their chemical constitution testifies, but the alteration in each case has not been so marked as to obliterate the resemblances between them. The fact that they are commonly associated in layers, which recall those exhibited by a section of a peat bog (Figs 8 and 9), strengthens this conclusion.

The Tertiary Lignites consist chiefly of Angio-spermous woods, as also is the case with the Forest beds associated with the Peat. Wood will also under certain circumstances pass into the state of lignite very rapidly, as may be sometimes seen in the timbers of old colliery workings. Count Solms-Laubach (38) states that in one mine timbers 150 years old were found to be converted into Lignites of black colour, and lustrous conchoidal fracture, while in another case as little as six years was necessary to produce similar changes.

Another point to be remembered is that nearly all Brown coals and Lignites, as well as Peats, have so far escaped the great pressures resulting from earth movements, to which most of the geologically older types of coal have been subjected. They have therefore only undergone changes chiefly of a biochemical nature (p. 134). Perhaps however the fact that many of the Brown coals and Lignites have been buried beneath an enormous weight of more recent sediments may have had some considerable influence in determining their present chemical composition, and the differences in physical characters which distinguish them from peats.

We now turn to the question whether the Humic, Sapropelic and Anthracitic coals of the Palaeozoic period have originated under terrestrial conditions similar to peat.

It is generally agreed that all coals have been formed under aqueous conditions of one sort or another. No case is known in which the evidence indicates that the mother substance has accumulated on really *dry* land. We have reviewed the conditions of formation of recent peats, and arrived at the conclusion that some, but not all, Tertiary Brown coals and Lignites may have had a similar origin. It is however quite certain that all coals have not been derived from peats, nor has the mother substance always accumulated under what we have here called terrestrial conditions. Recent studies of the structure of coal have tended to emphasise still further the conviction, which has been gaining ground for some years past, that coals have resulted from very dissimilar types of vegetation, and that the mode of accumulation of the mother substance has varied in different cases.

It must always be remembered that in the early days of the study of these vexed questions, only one process was known to be actually in operation, which furnished a picture of the conditions which might, in Carboniferous times, have led to the formation of coal. This was the growth of peat. The publication in 1845 of Lyell's *Travels in North America* (27), in which he described the conditions existing to-day in the Great Dismal Swamp of Virginia and North Carolina, and the accounts which have since appeared of other swamps in India and elsewhere, have furnished us with further material for restoring the conditions, which may have existed when the older coals were accumulated. More recently still, the work of Renault and Potonié has accustomed us to the view that lacustrine vegetation may have given rise to certain types of coals, especially the Sapropelic fuels.

The Peat-to-Anthracite Theory.

But to return to the question whether all coals did not originate under terrestrial conditions, similar to those of modern peats. It may be that what we will call the Peat-to-Anthracite theory still finds acceptance to-day. This hypothesis assumes that all coals 'started life,' so to speak, with an initial stage of Peat. The Peat later became converted into Brown coal, next Brown coal passed into Humic or Bituminous coal, and finally, at least in some cases, Anthracite resulted. The grounds for this view are firstly, that admittedly some Brown coals and Lignites have originated from Peats, as we have seen (p. 65); and secondly, that a table of the chemical composition of the various types of coal, such as that on p. 9, shows a progressive increase in the carbon ratios, and a similar decrease in the hydrogen and oxygen percentages, as we pass from wood or Peat to Anthracite, and suggests that a gradual transition of this nature may have taken place in the past. But it must be remembered that, however tempting such a table may prove, it by no means follows that such a scheme represents the sequence of changes which has taken place in nature. It has never been shown that Brown coal can be converted into Humic coal, still less that Anthracite has been actually derived

from the latter. Recent work on the origin of
Anthracite, as we shall see, p. 143, is entirely opposed
to this contention. If the Peat-to-Anthracite theory
holds, it is curious that we find no true Humic or
Anthracitic coals in the Cretaceous or Tertiary
periods.

The Peat-to-Anthracite theory involves the idea
of wide-spread terrestrial accumulations in Carboniferous times, resembling our modern peat-mosses.
It also implies, in addition to the prevalence of
climatic conditions suitable to peat formation, the
existence of a vegetation specially adapted to flourish
under such conditions. As we have seen (p. 63) these
two necessities do not always exist side by side at the
present day.

In the past several students of these problems
seem to have dissented from the Peat-to-Anthracite
hypothesis, on what now appear to be inadequate
grounds. It is admitted that peat formation to-day
is almost entirely confined to temperate regions in
the Northern hemisphere. Little peat occurs in the
tropics, except in Madagascar and possibly some
other localities. It has been assumed that, *because
the climate of the Carboniferous period was tropical
and not temperate*, therefore coals could not have
originated under conditions similar to those of modern
peats. This raises the vexed question of former
climates. So far as the writer is aware, however,

no evidence, based on a study of the vegetation,
has yet been brought forward, which demonstrates
emphatically that a tropical, rather than a temperate,
flora existed in Upper Carboniferous times. Recent
anatomical studies of fossil plants of this age have
only confirmed the impression that the difficulties
of arriving at any conclusions of value on such points
are at present almost overwhelming.

The arguments in favour of a tropical climate in
Upper Carboniferous times have been chiefly based
on the general character of the flora, and the supposed
identity (in part) of the classes of plants, then existing,
with those which still flourish to-day. The abundance
of ferns in Carboniferous times was instanced by
Brongniart and others as evidence that the climate
was then warm and moist. We know now that not
only was there no real predominance of this group
of plants, as Brongniart thought, but that the majority
of these plants, though fern-like in habit, were not
ferns at all, but seed plants (Pteridosperms). It is
of course quite impossible to infer from the distribution of modern ferns, what might have been the type
of habitat in which the Pteridosperms, a quite distinct
race, flourished in Carboniferous times, any more
than we could deduce from the habitats of living
ferns those of the Cycads, were the latter unknown
to us in the living state.

In brief, modern research has shown that the

Carboniferous flora was so unlike that which exists to-day, that any arguments from now to then are worse than useless, so far as evidence of climate is concerned. The fact that, in Carboniferous times, tree forms predominated and herbaceous plants appear to have been in a minority, does not seem to the writer to afford any special evidence on the subject of climates, when the enormous areas of primaeval forests, which still exist in the cooler portions of the temperate zone in Northern Europe and America, are called to mind. Another argument, namely, the large size of the cells of the majority of Carboniferous plants, which is such a striking feature of their anatomy, does not appear to have been a necessary corollary to a tropical climate, or to be adequately explained on such a hypothesis. Neither is luxuriance of vegetation confined to the Tropics. It may be seen even in an Alpine flora.

What we do know at the present time of the climate of the Carboniferous period is that it was extraordinarily uniform throughout the Northern hemisphere, and that the North Polar region then enjoyed a climate not markedly dissimilar from that of England and Southern Europe at the same period. The uniformity of the fossil flora of the Upper Carboniferous rocks of Europe and North America is extremely striking, as Lyell (27) long ago pointed out, and more recent work has merely tended to emphasise this conclusion, and to widen the boundaries of the single botanical province of that period in the Northern hemisphere by the inclusion of Central and Eastern Asia. To-day many floras, ranging from boreal through temperate to subtropical, inhabit the same area.

In connection with the question of the climate of the Carboniferous period, some discussion should perhaps be devoted to the theory that the atmosphere was then more highly charged with carbonic acid gas than it is to-day, and that to this fact was partly due the abundance of vegetation, much of which has since been converted into coal. The Carboniferous period has been described as the era of the purification of the atmosphere, when the excessive amount of carbon dioxide was withdrawn from the air by an abundant growth of tropical vegetation. In the ordinary way this carbon dioxide would have been returned to the atmosphere as the plants died and decayed, but owing to their conversion into coal, it became locked up in the rocks. In support of this contention, the absence of birds and the rarity of land animals at this period has been quoted, and it has been inferred that the atmosphere was not in a fit state for their existence, until the excess of carbon dioxide had been withdrawn from it.

Superficially the arguments may appear plausible, but many sound objections to the theory have been advanced by Lyell and others in the past, and it is extremely doubtful whether any such hypothesis can be justified. Lyell (27) has pointed out that at the present time the atmosphere receives large additions of carbon compounds during volcanic eruptions, and also from the decay of vegetable and animal matter, and other sources. 'But it does not follow,' he adds in his *Principles* (Vol. I, p. 277), 'that the air is becoming more and more loaded with carbonic acid, for there are causes in action which prevent such a change in the constitution of the atmosphere.'

The quantity of carbon dioxide to-day in the atmosphere varies from 3 to 6 parts in 10,000 measures, and this is sufficient to permit of coal formation on a considerable scale being in progress at the present time. It must also be remembered that the conditions have been favourable in various regions for coal formation during Mesozoic and Tertiary times. Thus, as Lyell adds, 'the abundance of coal, therefore, in certain districts may have arisen from the peculiarity of the vegetation, and of a climate which prevented decomposition, rather than from a peculiarity in the atmosphere which enveloped the globe in the Carboniferous period' (*ibid.* Vol. I, p. 228).

To return to the Peat-to-Anthracite theory. The most emphatic evidence which negatives this view, is that which indicates that Carboniferous coals were formed quickly. If such a coal as Anthracite

started in the condition of Peat, and then passed
through a Brown coal stage and next a Humic
stage, it is obvious that considerable periods of
time must have been necessary to allow these
changes to take place.   But there is strong evidence,
as we shall see in Chapter VI, that coal was not
only formed, but denuded, within a comparatively
short interval, and this fact in addition to the other
objections above indicated, has led to the rejection
of the Peat-to-Anthracite theory, and the substitution
of the metamorphic hypothesis of the origin of the
Humic, Sapropelic and Anthracitic types of fuel.
At the same time it is not denied that the initial
stage in the formation of such coals may have
resembled Peat in certain characters, or indeed that
true terrestrial accumulations may have given rise to
such fuels even in Carboniferous times.   Peat-like
accumulations or deposits may be formed under
a variety of circumstances, as we shall see in the
following chapter, other than those which we have
here termed terrestrial.   But, on the other hand, it is
not admitted that the mother substance has in all
cases accumulated under conditions exactly similar to
our modern peats, nor has the subsequent course
of events coincided with those postulated by the
Peat-to-Anthracite theory.

CHAPTER V

THE MODE OF FORMATION OF ESTUARINE
AND LACUSTRINE COALS

WE pass on now to consider how accumulation takes place under estuarine conditions. Under this term we include all coastal swamps, situated on or near the littoral, whether freshwater, brackish water, or marine. At the present day estuarine swamps are very varied, both as to their vegetation, and also as to the conditions of their existence. They are common along sea margins, and at the deltas of rivers, in both the Tropics and the Temperate zone. Freshwater swamps may exist inland, at some distance from the coast. In nearly all cases, such swamps are fed by streams or rivers, and, where they are not situated actually on the coast-line, they are also drained by streams running seawards. The material which accumulates in these swamps is partly or wholly derived from plants which grew actually in the swamp itself, and thus originates *in situ*. But in some cases, especially in delta swamps, a large amount of river-borne, drifted material may be added to the mass which has accumulated in place. A river carrying a considerable amount of drifted material will tend to deposit its load at or near the entrance to a swamp, because its velocity is greatly decreased by the swamp vegetation, which checks its flow. Thus the accumulation of vegetable debris formed under estuarine conditions may have originated partly from swamp vegetation, and partly from drifted material. For this reason the use of the term 'estuarine' appears preferable to 'swamp' in such considerations.

Modern Estuarine Accumulations.

We may now study the conditions determining the nature and existence of estuarine accumulations and deposits, in the hope that they may afford us pictures of the conditions under which the mother substance of certain types of Palaeozoic coals may have been laid down. As we have pointed out, estuarine deposits of the present day vary greatly. They differ from terrestrial accumulations in the fact that the waters in which they were laid down are free and circulating. We will begin with an account of a freshwater Swamp in a Temperate climate, in which the deposits have originated entirely *in situ.* The Great Dismal Swamp of Virginia and North

Carolina (United States), of which excellent accounts have been given by Lyell (27), and more recently by Shaler (37), is perhaps the best-known example of this class. (Fig. 10.)

The area of this swamp is rectangular, and, before considerable portions were reclaimed, it extended for about 40 miles from North to South, and 25 miles from East to West. It therefore exceeds in size many of our coalfields. A small lake, oval in shape, and about 7 miles along its greater dimension, exists more or less in the centre of the area. Its waters are very shallow, the maximum depth being estimated at between 6 and 15 feet. The swamp is fed by several rivers, and drained by streams running into the sea, some miles distant. It is not however affected by tides, but is wholly freshwater, and the waters are free and circulating.

The geological history of the swamp appears to have been as follows. It is a portion of the sea bottom, the surface consisting of Pliocene (p. x) sands, which has been raised and somewhat denuded, but which is now gradually sinking again (37). The seaward slope at present is not sufficient to carry off the waters, and the dense vegetation which covers the area further impedes their free course.

The waters of the swamp are quite clear, and free from sediment. The soil is black, and consists of vegetable debris, without any admixture of mineral

matter.   The swamp is thickly forested, the two
most important trees being the Swamp Cypress
(*Taxodium distichum*), and the White Cedar (*Cu-
pressus thyoides*) (Fig. 10).   The dead trunks of

Fig. 10.   A view of the vegetation of a temperate, freshwater swamp,
the Great Dismal Swamp of the United States.   After Shaler.

these trees are continually falling, and becoming
buried in the peaty soil.   Owing to the looseness of
the soil about their roots, they are easily overthrown

by the wind. The wood, covered up by the water, does not decay, but sinks into the peaty soil consisting of decayed roots, leaves, bark, branches and seeds. Lyell estimated the thickness of this soil at 10 to 15 feet. According to Shaler (37) the average rate of accumulation of such soils in North American swamps is $\frac{1}{10}$ inch in a year, or 8 feet in a thousand years. Thus, supposing such a soil became carbonised, and the volume of the coal was $\frac{1}{12}$ of the bulk of the 8 feet of soil, we should get a seam of 8 inches in thickness as the result of 1000 years' accumulation *in situ*, and the total present thickness of the soil would give us a seam only 15 inches thick.

Coastal, brackish-water and marine swamps are abundant in the Tropics. The *Nipa* and Mangrove swamps of India, Ceylon, and the Malay archipelago are examples (Fig. 11). The Mangroves grow in a black mud or ooze, their characteristic, branching stilt-roots, and the lower portions of their trunks being usually partly submerged in brackish or salt water, though these plants may also occur in freshwater swamps. The mud is a semi-marine deposit, which though in some cases originally fluviatile, is constantly being shifted by the action of the tides. All the dead and decaying vegetation of these highly forested swamps sinks into it, and the debris of the plankton, and the algae living attached to the Mangroves contribute to an accumulation *in situ*.

The *Nipa* palms, which form extensive dwarf forests, flourish abundantly in brackish water estuaries, where the water is much less salt than in the case of typical Mangrove swamps. The soil here also consists as a rule of a black ooze, no doubt chiefly of

Fig. 11. A view of the vegetation of a tropical estuarine swamp, a Mangrove swamp in Ceylon. From Tansley and Fritsch, after Lang.

fluviatile origin, and into it sinks all the debris of dead and decaying vegetation.

So far we have considered only vegetation which has accumulated *in situ*, in temperate and tropical regions. But it must be remembered that large

quantities of drifted materials are carried down
by most rivers in the Tropics, consisting partly of
aquatic plants growing in the river, and partly of
trunks and other debris, the result of the 'wear and
tear' of the vegetation which clothes its banks.   The
amount of this drifted material may be considerable
at certain seasons of the year, even in the case of an
English river.   The well-known rafts or floating
islands of the Mississippi and the Amazons, consisting of interlaced trunks of trees, sometimes with
other plants growing upon them, are other examples
on a larger scale.   These drifted materials may come
to rest in the estuary of a river, especially if a delta
swamp is in existence, and thus drifted debris may be
added to the mass of vegetable material, which has
originated *in situ* in the swamps of the delta.   In
times of flood, the amount of drifted material is
naturally greatly increased.

From this brief survey of recent, estuarine accumulations, we gather that they are very varied,
as to the place of origin of the material, as to the
nature of the material itself, and as to the conditions
under which it formerly flourished.

Palaeozoic Humic Coals.

We now turn to the question of the mode of
accumulation of Palaeozoic coals of the Humic type.
Did they originate under estuarine conditions at all
similar to those which are in existence to-day, such as
we have just discussed? If so, did the accumulation
of the mother substance of such coal seams result
from vegetation which grew on the spot, where the
seams are now found, or did it consist of debris
drifted from a distance? We will attempt to answer
the second question first. If some coals have
originated from drifted materials, then no doubt
the mother substance was deposited under estuarine
conditions. If other coals have been formed from
a mass of vegetable material, which once flourished
where we now find a coal seam, then such coals may
have originated under terrestrial conditions like Peat,
or under swamp conditions like those prevailing
to-day in the Great Dismal Swamp of North America,
or under lacustrine conditions, a further possibility
which we have yet to consider.

Both the 'drift' and the 'growth-in-place' theories
have been strongly urged in the past, each with
a view to showing that the alternative is, under any
circumstances, untenable. In fact, the chief controversy with regard to coal during the last hundred
years has always centred round these very questions.

To-day authorities are perhaps as much divided as ever they were. On the one hand Potonié (28) and his followers, while admitting that coals are of varied origins, believe that they have all been formed from growths *in situ*. On the other hand, the French school, especially Prof. Grand'Eury (16—18) and M. Fayol (10), have shown that some coals, at least in Central France, have originated from drifted vegetation. Some authors have recently concluded that, on the whole, the exponents of the Drift theory have the upper hand, while others maintain the contrary. The data are so complicated, and the evidence, if considered only in part, is so liable to mislead, that many writers, beyond a statement of the chief arguments on either side, have not attempted to express any personal opinion on these matters. It appears to have been overlooked that both theories may be equally true, that some coals may have been formed in one way, some in another, and that the circumstances which have existed in the past may have been quite as complicated, and quite as varied as they are to-day.

We will now discuss the chief arguments advanced on either side, but before doing so we will dismiss certain considerations, which may at first sight appear to bear on the matter, but which in reality have no special significance in one direction rather than another.

It is commonly observed that the plant impressions on a slab of shale from the roof of a coal seam, are both very fragmentary, and very mixed. Portions of pith-casts of *Calamites* may overlie fragments of fronds of *Alethopteris*, or *Neuropteris*, or both, and, in addition, the leaves of *Cordaites*, or *Sphenophyllum*, or the stems of *Lepidodendron*, or *Sigillaria*, may also be associated in a specimen less than a foot square. All the impressions are obviously broken. Among the petrified plants of the coal-balls (p. 106), which are found actually in the seam itself, we notice the same state of things. For instance a section, 5 × 3·5 inches, cut from such a concretion, may show fifteen or more fragments of *Lepidodendron*, *Calamites*, *Myeloxylon*, and Fern petioles, all contributing to a confused assemblage. From such cases it has been argued that we have an irresistible proof of the truth of the Drift theory. Nothing is really further from the case. Even the growth-*in-situ* theory implies that the material has accumulated in or under water, and not on dry land. It is extremely probable that, while in such cases the material actually grew on the spot, where we now find the plant-bearing shale or nodule, the vegetable debris was laid down under water, and mixed and re-sorted repeatedly by the action of currents or tides, without having travelled any distance. The Drift theory inherently implies transportation from a dis-

tance. Here, while a slight shuffling or rearrangement
may have taken place, there is no real evidence of
transportation. We shall return to the study of this
sorting action of water again, when we shall endeavour
to show that it is a conception of great importance
and value in connection with the problems with which
we are here face to face.

On the other hand, badly preserved impressions
of plants commonly occur in the sandstones and
shales associated with coal seams, which are undoubtedly of drifted origin. Occasionally, instead
of carbonaceous impressions, one finds, especially
in the sandstones, little, thin patches or streaks
of coal, which have resulted from a small mass
of drifted vegetation, which has become carbonised.
The material giving rise to these phenomena, corresponds to the sea-drift (Ger. *Häcksel*) found on
all coasts, both temperate and tropical. In England
it consists chiefly of seaweeds, but on tropical shores
the sea-drift is composed of a large number of
fragments of various plants, including twigs and
leaves, with fruits and seeds of Flowering Plants,
which are thrown up by the tides. It is important,
however, to bear in mind that, while such drifted
materials may occur in the rocks associated with
coal seams, this fact is no evidence whatever that
the coal seams themselves have resulted from an
accumulation of drifted matter. The conditions

under which a seam of coal has accumulated are quite distinct, as a rule, from those which prevailed when the overlying beds were deposited. The seam may in some cases have originated from estuarine vegetation, which grew on the spot where we now find the coal, whereas the overlying sandstones and shales were deposits derived from materials of marine origin.

The Growth-in-situ Theory.

We may now consider the chief evidence which supports the 'Growth-*in-situ*' theory, or autochthonous formation of coal, as it has been termed by Gümbel (21). This hypothesis postulates that the original mother substance of a coal seam accumulated in the place where the coal exists to-day, and that the material was not transported from a distance by means of running water. This is the older of the two views, and was founded originally on a comparison with peat-mosses, and with freshwater swamps.

The Purity of coal seams. One of the most remarkable features of a seam of coal, whatever its geological age may be, and in whatever part of the world it may occur, is its purity. By purity is implied the absence of any sand or mud mixed with the coal. If we contrast a coal seam with the associated rocks (Chapter II) in this respect, the purity of the seam is thrown into prominence. The associated sandstones are frequently impure. They pass laterally, as we have seen, into argillaceous rocks, with all gradations between the two. The shales are also frequently sandy, and may merge into arenaceous rocks. But a seam of coal consists of hydrocarbons pure and simple, and it is rare to find arenaceous or argillaceous impurities mixed with the coal. As is well known, it is the practice at most collieries to 'pick over' by hand the best coals, when they are brought up from the pit, and as they pass over the long travelling 'screens,' which lead from the shaft to the railway truck. The object of this process is simply to remove those lumps which contain iron pyrites, and also those to which adhere portions of the roof of the seam, or of the fireclay below. Otherwise the coal is pure.

The purity of coal has always been one of the great arguments in favour of the *in situ* theory, and rightly so. On the Drift hypothesis, it is difficult to explain why, if the original mother substance had been swept along for considerable distances in rivers, it did not become mixed with earthy particles, such as mud or sand. The ordinary laws of sedimentation no doubt apply here just as in other cases. It would almost seem at first sight that this consideration alone rules the Drift theory out of court in every instance. There is little doubt however that this is not so.

The original material may, in some cases, have drifted from a considerable distance, and it may have arrived in position in a highly contaminated state. But there is no reason to suppose, that when the material reached its present position, the process of accumulation at once ceased. It was still under the influence of water, though actual transportation had reached its furthest limits. It may have been re-sorted on the spot, under the influence of currents, which would sift out the sand and mud, according to their specific gravities, and leave an accumulation of pure, vegetable material.

We have evidence that the Carboniferous deposits were often affected by strong currents, as is shown by the frequent current-bedding of the sandstones, and an abundance of ripple marks. In the case of the great barren coalfield of Devonshire, which occupies the greater portion of that large county, and also part of North Cornwall, it would seem that the absence of coal, with the exception of a few seams in the neighbourhood of Bideford, is well accounted for by the evidence of the existence of strong currents in this area in Upper Carboniferous times. These swift streams swept away the accumulating vegetable materials, and scattered them far and wide, and consequently to-day the coalfield is practically barren. We may therefore conclude that the sorting and shifting action of currents, large and small, has played a part, at least in some cases, especially where thin coals are now found, in the preparation of the accumulated vegetable material, before its actual conversion into coal.

The essentially modern view that certain types of coal have originated from accumulations of one particular class of plants, is another argument in favour of the *in situ* theory. As Lindley and Hutton in their *Fossil Flora* (Vol. II, p. xxvii) remarked, as long ago as 1833, 'if the vegetables had been washed from a distance, is it likely that the different kinds should have separated so completely, as to have produced the several varieties of coal so distinct from each other?' In this connection it may be argued that many coals have resulted from extremely mixed assemblages, representing almost all classes of Carboniferous vegetation, while in other cases there appears to be evidence that the coal has originated from one particular plant type. The growth-*in-situ* hypothesis may be true of the latter case, but it does not follow that it is always true of the former.

The Uniformity of Thickness and Composition of Coal Seams.

The supposed uniformity of thickness and composition of coal seams has been urged as an argument in favour of the *in situ* theory exclusively. But it may be easily dismissed, for the exceptions are more numerous than the rule. One seam, often the principal seam of thick coal in a district, may be fairly uniform in thickness and composition, but the instances, especially among the smaller and less important seams from an economic standpoint, where this is certainly not true, may be numbered by hundreds or even thousands. Those who have ever attempted to correlate the vertical sections of the twenty or more seams passed through in three or four shafts, placed along the strike of the same coalfield, will be under no delusions on this point. The larger seams and some others, the floors or roofs of which are characterised by some special but local peculiarity, may be often linked together, but it will be found that a whole group of coals seen in one shaft is absent in a neighbouring sinking, and in practice the geologist is somewhat fortunate if he can establish a general, though not detailed, correlation. Neither is the composition constant, as we have seen (p. 18). A Humic coal may pass into an

Anthracite, a Cannel into a Humic coal. There is then clearly no proof of the *in situ*, rather than the 'drift' hypothesis, or *vice versa*, to be found in the evidence relating to the thickness and uniformity of composition of coal seams.

The Underclays. We have already called attention (p. 27) to the unstratified clay deposits which may occur immediately below coal seams, and which are perhaps most widely known to geologists as *underclays*. In recent years these beds have attracted great attention, especially from the point of view of the formation of coal, and the literature on the subject is fast becoming enormous. The explanation of the origin of these clays as being *old soils* is no doubt the chief prop of the *in situ* theory, and while it may be conceded that, in many cases, the seams above them have originated *in situ*, it must not be imagined, as has so frequently been assumed, that the presence of an underclay invalidates the drift theory in other cases.

The significance of the occurrence of underclays beneath coal seams was apparently first pointed out by Logan (26) in 1840, though no doubt the miner had been perfectly familiar with them for centuries previously. Logan showed that in South Wales, such clays always underlie the coals, though coal seams do not always occur above every bed of fireclay. These clays are full of the rhizophores of *Lepidodendron, Sigillaria*, and other Palaeozoic

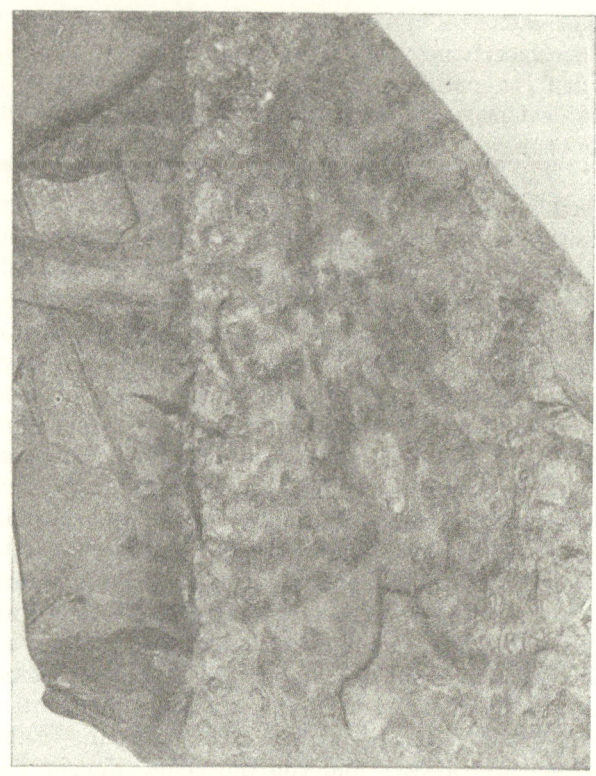

Fig. 12. Part of a rhizophore of *Stigmaria ficoides* (Sternberg), showing the scars of attachment of roots. Roots still attached are seen on the left-hand side. From the Middle Coal Measures of Canonbie. No. 1083, Carbon. Plant Coll. Sedg. Mus. Camb.

Lycopods, to which are frequently still attached numerous, lateral roots. These roots and rootlets also commonly occur detached. These rhizophores are known under the name *Stigmaria ficoides* (Sternberg) (Fig. 12) and both their internal and external morphology has been very thoroughly worked out long since. A few other rare species are known, but, so far as impressions are concerned, ninety-nine examples out of a hundred conform to the type first described by Sternberg. In other words the rhizophores and roots of all Carboniferous Lycopods were so much alike, that we cannot distinguish those of a *Lepidodendron* from those of a *Sigillaria*. These rhizophores and roots are also abundant in the calcareous nodules, known as coal-balls (p. 106), where the structure of the enclosed plants is preserved.

Stigmaria is by no means confined to the underclays (Fig. 2), but is also frequent in the shales and sandstones associated with coal seams in both the Upper and Lower Carboniferous rocks. The flora of the underclays however consists as a rule, though not always, exclusively of *Stigmaria*. So far as the writer is aware, the roots of other classes of plants have not been recognised in the underclays.

Logan assumed that, because, in South Wales, underclays are always to be found below the seams, they invariably occur in this position elsewhere, and that they represent the old soils on which the vegetation, now converted into coal, once grew. The *Stigmarias* are the underground portions of these plants still *in situ*. This conclusion is still current among geologists. Recently attempts have been made to show that underclays occur below coals of Jurassic age (14).

Before we pass to criticise the conclusions drawn from this evidence, we may further point out that in the underclays, in addition to *Stigmaria*, there sometimes occur stumps or bases of trunks, in rare cases with Stigmarian roots attached. Short portions of the trunks of trees, either recumbent or upright, are also found in the rocks associated with coal seams in certain districts, and the evidence to be drawn from their occurrence has given rise for many years past to a heated controversy, which still rages fiercely. We shall return to this subject shortly, but for the present we will concern ourselves with the evidence of the underclays and their contents.

It was a facile assumption, before the subject had been closely studied, that those stumps, occurring in the underclays, were in all cases *in situ*. These, it was argued, are the stumps of the trees, still rooted in the old surface-soils, the upper parts of which have vanished, for the simple reason that they have contributed to the bulk of the coal, which now lies above them. This may well be true in certain cases, but it does not follow that the presence of the bases of trees in the underclay is evidence that the coal has originated from material accumulated *in situ*.

There are certain difficulties in connection with the evidence of underclays, which we will now discuss. The first of these is their manner of formation. What do they really represent? They are commonly claimed to be old surface soils, but if this is true they are certainly not terrestrial deposits. Nothing could be more unlike a soil, in the usual sense of the term, than an underclay. But if we admit that they are of aqueous origin, what precisely were the circumstances of their deposition?

The writer has inquired of Mr Lewis, whether he finds that the modern peats rest on clays, and he is informed that, while in some districts the peat may be underlain by clays or stratified muds, this is not always the case, and the occurrence of these beds is found to vary with the geology of the district. On the other hand, where the peat is thin, the roots of the trunks of the Upper Forestian (see p. 60) are sometimes found to penetrate the sands and clays lying below this horizon. Beds of clay may also sometimes be interstratified with the layers of peat.

On the whole, it is not evident at present that

any strict parallel exists between the coal seams with their underclays and the *layers* of peat, even when the latter rest on clay beds, which appears to be a somewhat exceptional circumstance. When we turn to the freshwater swamps, such as the Great Dismal in North America, we do not find any evidence at all of clay beds below the peat, which rests on Pliocene sands. Lyell (27), p. 146, expressly states that 'the black soil...to which the mosses and the leaves make annual additions, does not perfectly resemble the peat of Europe, most of the plants being so decayed as to leave little more than a soft black mud, without any traces of organization.' Since this soil is formed of 'vegetable matter, usually without any admixture of earthy particles' (*ibid.* p. 145), it has obviously nothing in common with underclays.

It would appear that the underclays of our coal seams more closely correspond to the black oozes of marine and semi-marine estuarine deposits, such as those of the Mangrove and *Nipa* swamps of the Tropics, which we have already discussed (p. 80). In such muds, roots and stumps of trees occur to-day *in situ*, though it must be remembered that drifted material of a similar nature may also become buried in the ooze. The muds surrounding the stumps of trees in the buried forests of our coast-lines offer another parallel.

In the author's experience, an underclay is typically a hardened unstratified mud. In Belgium, Dr Renier (32) finds that there is no real lithological difference between an underclay and a roof shale, and that the physical characters of the former are due to the fact that they are permeated by casts of *Stigmaria*, which determine the planes in which they split. However this may be in Belgium, the present writer is unable to support Dr Renier, so far as British underclays are concerned. They are physically distinct from the roof shales as well as unstratified. The differences between the two types are so well marked that, even where fossils are absent from both, no one could mistake the one for the other.

If we agree that the nearest analogy to the underclays of our coal seams is to be found in the marine estuarine oozes of tropical coasts, it must be remembered that such deposits are being constantly shifted and re-sorted, under the action of tides and currents to-day, and this applies also to the vegetation buried in them. It therefore does not follow that all the *Stigmaria* found in the underclays are *in situ*. The fact that casts of the rhizophores commonly occur, without any rootlets attached, indicates that drifted material may have become mixed with relics of the vegetation which flourished in place.

Although, as Logan (26) pointed out, underclays occur beneath all the coals in South Wales,

A.   7

and as Lyell (27) showed, the same is true of the coalfields of the United States and Canada, the known exceptions ro this rule are now numerous. Not only are fireclays commonly found without any coal seams above them, but they may occur as the roof above the seam, or in the seam itself. Where they are absent below the coals, the fact may perhaps be accounted for on the supposition that the accumulating vegetable material has been swept away by currents (see p. 88). But sometimes coals occur without any underclay, and rest directly on sandstones, limestones, conglomerates or even on igneous rocks. Such coals may hare been formed under lacustrine, rather than estuarine conditions.

Another difficulty in connection with underclays is that their thickness commonly bears no relation to the extent of the sean above. Often thick coals overlie thin underclays, and *vice versa*. Too much weight cannot however be attached to this fact, for it is evident that, in the process of carbonisation, the mass of the vegetable aceumulation has been enormously contracted, and not always to a like degree. Further, where thick underclays and thin seams occur, it is difficult to determine, whether or no, part of the accumulating vegetable material may not have been swept away by currents.

The occasional occurrence of fragments of leaves of *Neuropteris* or *Alethopteris*, in association with

Stigmaria, is not uncommon in the underclays, and worn pebbles of quartz or other rocks, or fragments of shells, may also be found. These facts have no special significance in favour of the 'Drift,' rather than the *in situ* hypothesis. We should expect to meet with similar fragments occasionally in a modern estuarine ooze.

Upright trunks in the Coal Measures. As has been already stated, upright trunks occur, not only in the underclays, but in the sandstones and other rocks associated with the coal. These trunks are abundant in a few localities, though they are rare when the Coal Measures are considered as a whole, so rare indeed that, where they have been found to occur, they have usually been figured as being exceptionally interesting. They were described from the South Lancashire coalfield as early as 1839. Dawson (6) and Lyell (27) have discussed a remarkably interesting series of upright trunks at South Joggins in Nova Scotia. Grand'Eury (15) has described many examples from the coalfields of St Étienne and Gard in Central France, and, more recently still, Schmitz (34) and Renier (32) have figured other interesting examples from the Belgian coalfields.

In some cases the trunks are so numerous as to suggest that we are dealing with a fossil forest. Goldenberg long ago described a *Sigillaria* forest from the neighbourhood of Saarbrücken in Germany. One of the best examples in this country is preserved in the Victoria Park, Glasgow, and has been figured by Seward and others. It is of Lower Carboniferous age.

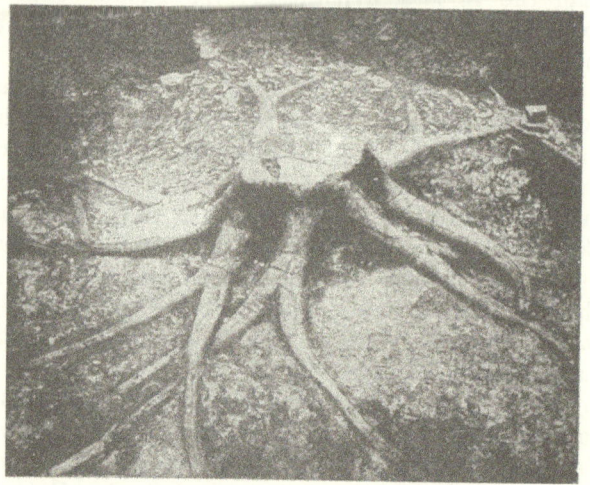

Fig. 13. A large example of *Stigmaria ficoides* (Sternberg) at Clayton, near Bradford, now in the Manchester Museum. After Williamson.

In addition, the bases of trunks, often on a large scale and quite upright, may occur in the underclays. A fine example, now preserved in the Manchester Museum, from Bradford in Yorkshire (Fig. 13), shows the four large, bifurcating arms of the Stigmarian rhizophores, still in continuity with the stem. Another similar fossil mounted in the Museum of the School of Mines in Berlin, was originally obtained from Piesberg near Osnabrück. These two specimens are the largest examples of fossil plants at present known.

The occurrence of upright trunks or stumps of trees in the Coal Measures has been regarded as good evidence of growth *in situ*. It is however a much disputed point whether this is always a safe conclusion or not. The circumstances existing in one locality are not always precisely similar to those of another; a fact which we must bear in mind in attempting to draw general conclusions from evidence of this nature. In other words, it is quite possible that the stumps observed in one district may be *in situ*, whereas those in another may be drifted material.

Let us examine some instances more closely. In Fig. 2, p. 28, three stumps, quite rootless, are represented in a bed of sandstone. One occurs immediately above a coal seam, the others are enveloped in sandstone at a higher level. It is not certain that these trunks, despite their upright position, are *in situ*. They are much more probably drifted material, like the sandstones which enclose them. It must be remembered that a drifted trunk will often remain upright, for its centre of gravity is low, and is situated at the thickened base of the trunk.

Figs. 14 and 15 are taken from Lyell's description of the trunks at South Joggins, Nova Scotia (27). Fig. 14 shows the mode of occurrence of five trunks in relation to the seams and other strata. Fig. 15

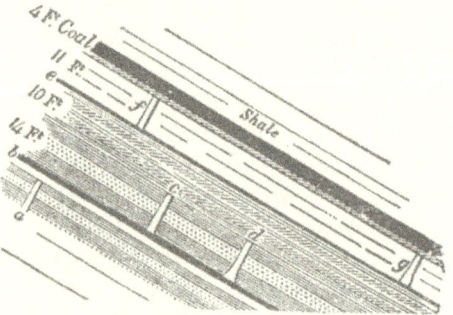

Fig. 14. Diagrammatic section of the Coal Measures of South Joggins, Nova Scotia, showing the position of five erect trunks, *a*, *c*, *d*, *f* and *g*, in relation to the coal seams above and below (the 4 foot coal, and the thin seams *e* and *b*). After Lyell.

is particularly important. No part of the original plant is preserved, except a cast of the bark. The stem was at one time hollow, and the cavity included in the bark has become filled up with sediments. The important point to be noticed is that the nature and order of the sediments filling

the hollow trunk, as seen in Fig. 15, do not agree
with those surrounding the trunk.   Lyell and others
explain this fact by assuming that, at the time when
the hollow stem existed, the waters were in a more or
less turbid state and by 'various other accidents,' and
do not doubt that such fossils were actually in place

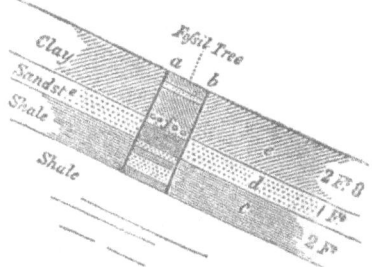

Fig. 15.   Diagrammatic section of the strata within an upright trunk
   a, b, in relation to the strata enclosing the trunk (South Joggins,
   Nova Scotia).   N.B.   The beds are tilted in this region, with
   a dip of 24°, but the axis of the trunk is perpendicular to the
   stratification.   After Lyell.

of growth.   The present writer is, however, in sym-
pathy with the view that it is quite as likely, if
not more probable, that the hollow stems of such
trees were partly filled with sediments, when the
trunk was prone, and, in that case, the evidence that
such fossils are *in situ* is not conclusive.   It has been
pointed out, some years ago, that the comparative

rarity of such examples of upright trunks is remarkable, if the *in situ* theory holds, for we should expect to find their bases occurring abundantly, rooted in the underclays. Gresley (20) has gone so far as to say that no *bona fide* case has yet been described where an erect stem has been found with its Stigmarian rhizophores attached, and rooted in the underclay, while the stem itself passed into the seam above.

In the case of certain trunks occurring in the Belgian coalfields, in an upright position in the roof of the seam, it has been found that a flattened cast of another plant, such as a *Sigillaria*, may occur between the base of the trunk of a Calamite, and the coal below it (9). In such cases the lower portion of the erect Calamite has obviously not contributed to the bulk of the coal which lies beneath it.

In some instances no doubt these erect stems are *in situ*, and the evidence is always strong in this direction, when the roots are found still in continuity with the trunk. A particularly clear case has been cited by Grand'Eury from the coalfield of Gard (France). Here a number of erect trunks of *Calamites* occur (Fig. 16) showing the successive formation of roots from higher and higher nodes. No doubt the explanation here is that the soil, on which they grew, became gradually silted up with sediments, and thus fresh roots were formed from higher nodes as they were needed.

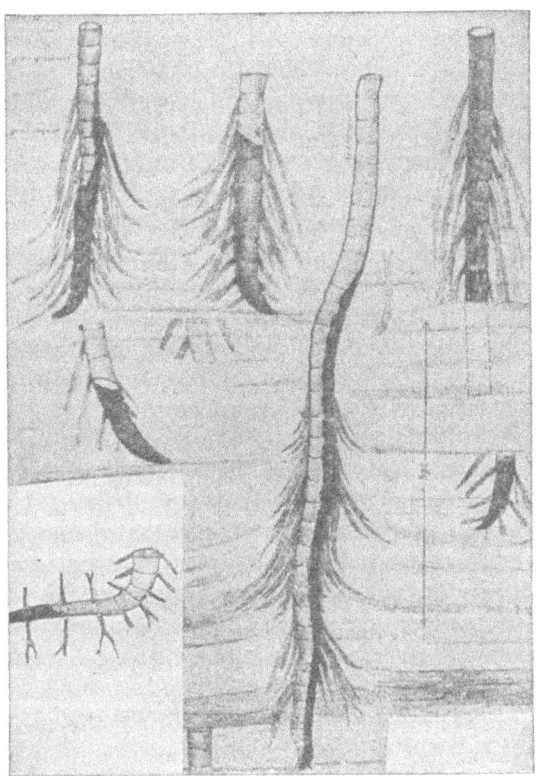

Fig. 16. Stems of *Calamites*, showing roots developed successively at higher and higher nodes as the soil was silted up around them. From Potonié after Grand'Eury.

It is usually found, especially in the roofs of seams containing trunks, that the erect stems are less numerous than those which are prone, or at any rate not upright. But even this evidence, if unsupported, is not satisfactory as an indication of drifted conditions, for such a state of affairs might occur in a water-logged forest, which is being drowned by the encroachment of a swamp (Fig. 18).

The upright trunks met with in the Coal Measures are almost always cut off square above, and not fractured obliquely as often occurs to-day in forests where a tree has fallen and broken off above the roots. It is however very difficult to decide as to what inference may be drawn from this fact, and indeed the whole question of the evidence of upright trunks in the Coal Measures is extremely perplexing, when it is remembered that, in the deltas of large rivers to-day, the bases of trees of large size, sometimes with fragments of roots still attached, may be deposited in a more or less upright position, though they have been transported for a considerable distance (10).

Coal-balls or *Plant-bearing Calcareous Nodules.* We may now pass from a consideration of the main arguments, brought forward to support the *in situ* theory, to a special case, where the evidence appears to be conclusively in favour of *in situ* accumulation. The Lower Coal Measures of Yorkshire and Lancashire, which at one time formed a continuous sheet,

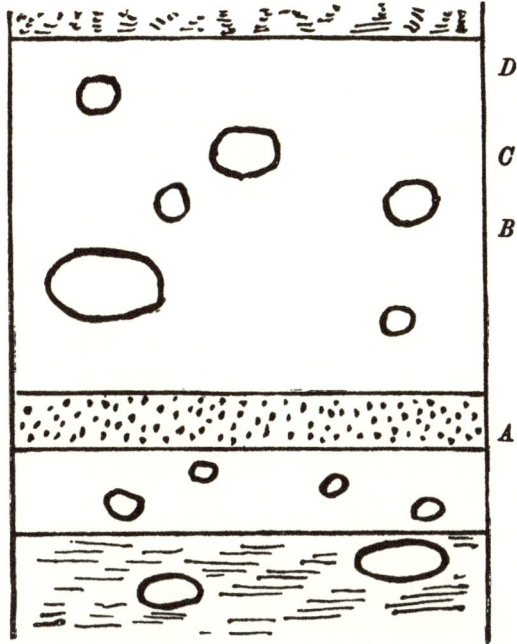

Fig 17. A section of the coal seams bearing calcareous nodules, which contain plant petrifactions, in South Lancashire. *A*. The Gannister coal (4 ft. 8 ins.) containing nodules and resting on its underclay. *B*. Sediments (8 inches) separating the two coal seams. *C*. The Upper Foot coal (1 foot) with nodules. *D*. Shale roof of Upper Foot coal, with nodules containing drifted plant remains, and a marine fauna. After Stopes and Watson.

stretching right across England, but which have since been thrown into two basins, on either side of the Pennine Chain, as the result of great earth movements, are famous for the plant-bearing nodules which they contain. These 'coal-balls,' as they are often called, are found nowhere else in Britain, and but very rarely in Continental Europe, and further they are confined to two seams, known as the Gannister and Upper Foot coals in Lancashire, and their equivalents in Yorkshire. The 'coal-balls' are not, as their unfortunate name implies, balls of coal, but calcareous concretions, in many respects quite like the other calcareous nodules, which abound in the Coal Measures (see p. 30). These particular nodules however are remarkable for the fact that they contain well-preserved plant petrifactions, from which the anatomy of the plants themselves can actually be studied. They have thus proved of priceless value in advancing our knowledge of the botany of the Carboniferous period. In recent years these seams in Lancashire have been worked entirely for the sake of their nodules, and for purposes of scientific research.

A section showing the distribution of the nodules is given in Fig. 17. It will be seen that the nodules occur both in A and C, the Gannister and Upper Foot coals, and that they are also found in the roof, D, of the latter seam. These roof nodules are however entirely different from the seam nodules, both in their flora, and in the associated fauna of marine *Goniatites*. While the seam nodules represent vegetation accumulated *in situ*, the roof nodules were no doubt formed around drifted vegetation.

Stopes and Watson (39) have recently completed a special study of the mode of occurrence and origin of these plant-bearing nodules. It will not be necessary here to enter into these matters, except in so far as the work of these authors bears on the mode of origin of the coal seams. The fact that the petrified plant-remains in the seam nodules, which these authors show were formed *in situ*, are extremely well preserved, indicates that the material forming the seam, which is obviously a pure accumulation of vegetable debris, has not been transported, but has accumulated in place. The striking contrast in the preservation of the plant-remains in the roof, in the marine nodules, as compared with those of the seam, quite apart even from their association with a marine fauna, is sufficient to indicate that both the shales forming the roof of the Upper Foot coal, and the plants, which later became contained in the nodules, were derived from drifted sediments of marine origin. We have here also a striking illustration of the fact, on which omphasis has already been laid, that the history of the roof of a seam is no criterion of that of the coal beneath it.

The Drift Theory.

We now turn to the alternative hypothesis of the Drift theory or Allochthonous formation of coal as Gümbel (21) termed it. This is the view which was specially advocated by Link (25) in 1838. It is a theory entirely independent of the conclusions derived from a study of underclays. It denies that the Palaeozoic coals originated under terrestrial conditions, similar to a modern peat-moss, or to the deposits now accumulating *in situ* in freshwater swamps, such as the Great Dismal. The material, which formed the mother substance of coal seams, was, on this view, always drifted into position, and transported from a distance (17, 18).

The arguments brought forward in support of the Drift theory consist partly of criticisms on, or the rejection of, the conclusions drawn from the facts advanced by the opposite camp, and partly of other details, which appear to bear out the view in question. In some cases, the conclusions appear to be based on somewhat superficial reasoning. Jukes, for instance, in 1849 argued that since the great mass of the Coal Measures is obviously water-borne and water deposited, it would be curious if the coals, which are interlaced with the other sediments, were not also derived from a deposition of sediment in water. As we have seen, this is certainly the case, but it by no means follows that the mother substance was accumulated from drifted materials. Further, the history of the various rocks associated with coal is so diverse, that there are no good grounds for the assumption that one litho-logical type has resulted under similar conditions to another.

According to the exponents of the Drift theory, the areas occupied by coal-bearing rocks are too large to have originated under estuarine conditions such as exist to-day. The force of this argument is emphasised when we remember that the Coal Measures once formed continuous sheets over almost the whole of large countries, such as England, where to-day we find only a number of isolated basins. Not only have the Carboniferous rocks been greatly denuded since their deposition, but they have been thrown into ridges and hollows, as the result of earth movements. How great the subsequent denudation has been is shown by the fact that the beds have been almost entirely removed from the ridges, and our present coalfields, representing the hollows, have alone sur-vived, owing to their sheltered position.

Again, coal is usually a stratified rock, and this is what we should expect if it were derived from drifted material. It is possible however that this stratification may, in part at least, be due to the heterogeneous nature of the vegetation forming the mother substance.

Another argument urged by the Drift theorists is that the phenomenon known as the 'splitting' of coal seams is difficult to account for on the growth-in-place hypothesis. It sometimes happens that a seam of coal, as traced laterally, will be found to split into several seams, separated by wedge-like intercalations of shale or sandstone, which may vary in thickness from an inch to many feet. These beds are known to the miners as 'partings' or 'dirt beds.' One of the best examples of an important seam splitting into several coals is the Thick or Ten-yard seam of the South Staffordshire basin, which when traced northwards splits into 10 to 14 distinct seams. In the southern portion of the Warwickshire coalfield also there is only a single thick seam of any economic importance, and this splits into five or more, widely separated coals to the north of the field.

The splitting of coal seams is probably more frequent than is generally imagined, for it is only where the main coals of the district are affected that the fact becomes of economic importance. No doubt the explanation is to be sought for in the oscillations of the earth's surface in Carboniferous times, when, as we have seen, there were alternate periods of elevation and depression. In South Staffordshire and Warwickshire, instead of an even depression, the ground appears to have undergone a local tilt, the southern area being somewhat raised, and the northern depressed.

Thus wedge-like encroachments of marine sediments overwhelmed the accumulating vegetable material unevenly to the north, while the raised southern portion of the seam escaped.   A series of such depressions and elevations following one another, of a like nature and constant axis, would appear to explain the facts presented by the splitting of coal seams, and, if this is the case, the matter has no particular bearing on the truth of either theory.

Another phenomenon of interest in this connection is that known to miners as 'rolls' or 'swells,' where the floor of the seam, whether of sandstones or other lithological types, instead of being level, is rolled into long swells or wave-like forms, the crests of which may project above the coals which occupy the hollows. This is sometimes regarded as an argument in favour of the drift theory.   In most cases these 'rolls' or 'swells' are no doubt simply irregularities in the level of the floor, which existed prior to the deposition of the seams.   If, however, we admit the existence of coals formed under lacustrine conditions, then such circumstances appear to offer no particular difficulty on the *in situ* hypothesis, though in some cases, naturally, the material may have been drifted into the hollows of an old denuded land surface.

The evidence to be drawn from upright stems, which we have already fully discussed, is another great argument of the Drift exponents.   As we have seen these stems in some cases are certainly not *in situ*. Examples have been found which are upside down, and in some districts the prone stems far exceed those still upright. No doubt the majority, if not all of these trunks, have been drifted. These stems never, or at least very rarely, occur in or traverse coal seams. Hence, while their presence may prove that the sandstones or other rocks in which they are included have been derived from drifted materials, it does not follow, as we have had repeated occasion to point out, that the mother substance of the seams was also formed of drifted matter.

Palaeozoic Estuarine Deposition.

We have now reviewed the chief arguments which have been advanced in support of the 'Drift' and '*in situ*' theories. We must next attempt to picture the conditions which were characteristic of estuarine deposition in Palaeozoic times. For those who hold to the *in situ* hypothesis, pure and simple, the Great Dismal Swamp of North America offers the nearest parallel existing to-day on the one hand, and the modern peat beds of Northern Europe, a possible alternative on the other. It must be also borne in mind that swamp conditions, though distinct from those of a peat-moss, on account of their free, circulating waters, may furnish essentially similar products. In cases where coals have originated *in situ*, it must be often very difficult to decide whether the conditions were originally those of a freshwater swamp, or of a peat-moss.

The Great Dismal Swamp, it will be remembered, is thickly forested with a Temperate flora, and gives rise to freshwater accumulations, derived entirely from the vegetation which grew *in situ*. As Lyell [(27) p. 149] remarks, 'Huge swamps in a rainy climate, standing above the level of the surrounding firm land, and supporting a dense forest, may have spread far and wide, invading the plains, like some European peat-mosses when they burst; and the frequent submergence of these masses of vegetable matter beneath seas or estuaries, as often as the land sunk down during subterranean movements, may have given rise to the deposition of strata of mud, sand, or limestone, immediately upon the vegetable matter. The conversion of successive surfaces into dry land, where other swamps supporting trees may have formed, might give origin to a continued series of coal-measures of great thickness.'

A somewhat different explanation of the origin of estuarine coals, based on the belief that the material may have originated partly *in situ*, and partly from drifted vegetation, is as follows.

It is now admitted that the Upper Carboniferous period was one of 'ups and downs,' that is to say, the surface of the land was either slowly sinking, or being gradually raised. An enormous number of alternating periods of elevation and depression appear to have taken place. During the periods of elevation, the land was thickly forested by a vegetation which was of a varied nature, but in which tree forms predominated in every class. There is no reason to believe that such typical representatives of the classes Lycopodiales, Cordaitales and Pteridosper-meae as *Lepidodendron*, *Cordaites* and *Lygino-dendron* were confined to swamp habitats.

The *Calamites* (Equisetales) may have been chiefly marsh-loving plants, and *Lepidodendron* and *Sigillaria* may have also flourished under similar conditions, but there are at present (22) no grounds for the supposition that any of these genera were *confined* to swamp habitats, still less to salt-water marshes, any more than the Mangrove, *Rhizophora*, is restricted to salt-water swamps today. It is more natural to assume, in the absence of indications to the contrary, that the ecological formations of Car-boniferous times may have been little less varied than those existing to-day in a Temperate region, such as Western Europe. Undoubtedly there were then as now, varied upland associations (22), as well as hygrophytic and halophytic, lowland assemblages.

No doubt, while the periods of elevation were in progress, estuarine conditions whether freshwater brackish-water, or marine, existed in Carboniferous times as to-day. But when the land began to sink, two changes took place. The sea invaded the estuaries and swamps, which became salt or brackish-water, instead of, as formerly, brackish or freshwater. Secondly, the swamps invaded those portions of the land which were formerly high ground, but had now begun to sink. The result was that large areas of almost level surface sank to near sea-level or even below, and the boundaries of the future coalfield were initiated. The dimensions of the existing delta and freshwater swamps became greatly magnified, and finally they merged into one vast swamp, and on this area vegetable debris accumulated.

The process of subsidence was however exceedingly slow, and plants continued to flourish for a considerable period over the whole of the slowly sinking district. The debris, which accumulated on the floor of the submerged surface, was almost entirely derived from vegetation, for owing to the sinking of the land, the streams would deposit their loads of earthy detritus far inland, and not as formerly at their mouths, now submerged. Thus a pure accumulation of vegetable origin would result, just as is found to-day in the Great Dismal Swamp. This accumulation would result, partly from the former vegetation which grew on the area, and was essentially of an estuarine character, and partly from plants which still forested the

sinking land. But, in addition, the vegetation of the
higher lands, formerly far removed from the estuaries,
became involved. It is well known that, when a swamp
at the present day invades a forested district, the trees
can usually no longer flourish in a water-logged soil
(Fig. 18). This no doubt was also the case in the
past, with the result that plants, which did not grow
under estuarine conditions, contributed to the mass

Fig. 18. Diagram showing the progressive stages of invasion of
a forest by a swamp. A A A, stumps of trees preserved in the
swamp; B B B, dead trees with branches standing; C C, forest
stunted by swamp growth. After Shaler.

of the vegetable accumulation. Thus the varied
nature of the flora of a coal seam may be easily
explained.

We have good evidence from fossil plants of the
stages in the gradual submergence of a forested area
in Carboniferous times. We have already discussed
(p. 104 and Fig. 16) the case of upright Calamite stems
which, as the soil became silted up, developed fresh
roots from higher and higher nodes. There are also

data, as Dawson and Barrois (1) have pointed out, which tend to show that the plants of the period

Fig. 19. Casts of the worm *Spirorbis* on a frond of *Sphenopteris neuropteroides* (Boulay), from the Upper Coal Measures of the Forest of Dean. No. 1733, Carbon. Plant Coll. Sedg. Mus. Camb.

may have existed for a considerable time after complete submergence. The leaf of *Sphenopteris neurop-*

teroides (Boulay), for instance, shown in Fig. 19, exhibits on almost every lobe of the pinnules or leaflets the characteristic casts of *Spirorbis*, which was an aquatic worm living in clear water. Such specimens indicate that the water, which overwhelmed the vegetation of the sinking land, was relatively pure and free from sediment, and that the submerged vegetation continued alive for a sufficiently long period to permit of the formation of these worm casts.

So far, we have considered only the origin of such material as grew *in situ*. But from those portions of the area, which formerly represented high ground, the rivers and streams still continued to pour into the estuary as large an amount of drifted material as before. Derived debris will have also been deposited at the seaward end of the swamp before the land began to sink, at or near, what were originally the mouths of the streams. Supposing the depression to have had an oval form, we should expect to find that while the accumulation of vegetable debris towards its centre might have resulted from plants which grew on the area, yet the deposits near the margin would have been derived, partly from drifted plants, and partly from vegetation *in situ*. As however the size of the depression was continually increasing, the admixture of the material was no doubt constant over the whole area.

Such a picture as this will explain the varied

nature of the vegetable material, which gave rise to a single coal seam, a conclusion of great importance as we shall see in the following chapter. It is also evident that the occurrence of underclays, perhaps corresponding to the black oozes of a Mangrove swamp, is not evidence that the whole of the material, which gave rise to a coal seam, was necessarily derived entirely from vegetation *in situ.*

But to follow the changes still further. The gradual nature of the subsidence allowed a growth and accumulation of vegetable origin to continue for a considerable period. At length, however, a stage was reached, when mud or sands, probably of marine origin, invaded the area, and covered over the accumulation of vegetable debris. The last remnants of the vegetation, now submerged, were swept away, though traces of them, enclosed in the sediments, which later formed the roof of the coal seams, have come down to us.

The area next entered on a new phase. A considerable thickness of sandstone, limestone or shale, purely marine or littoral sediments, containing drifted fragments of the vegetation of the neighbouring land, were deposited, and then the conditions changed again. The area was now gradually upheaved, and, as the circumstances became more and more favourable for the existence of plant life on its surface, it was invaded by the plants of the adjacent land, which had not undergone submergence. Thus a new cycle of *up-heaval—abundant vegetation—depression—carbona-ceous deposits*—was initiated, and this sequence was repeated hundreds of times during the Carboniferous period.

It appears to the writer that such a picture, necessarily imperfect, will explain a large number of the facts of Coal Measure geology, and reconcile the chief arguments of the *in situ* and Drift theories.

It must also be remembered that some of the Lignites or Brown coals of Tertiary, or even Creta-ceous age, may have originated under estuarine conditions, either from material accumulated *in situ*, or transported from a distance. The ultimate product of the Great Dismal Swamp in the next era will no doubt be a thick bed of Brown coal, intercalated with layers of Lignite, both *in situ*. In marine swamps, however, drifted material does occur, and this will eventually form similar deposits, but of a different origin.

Lacustrine Accumulations.

There is a further set of conditions, quite distinct from those of estuaries or swamps, under which a considerable mass of vegetable material may ac-cumulate to-day. Potonié (28) and others, in recent years, have made a special study of the deposits of

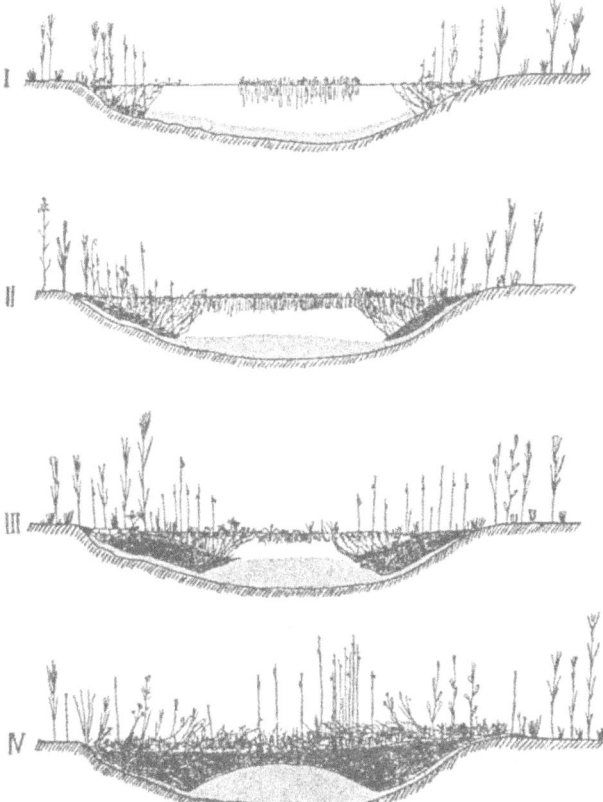

Fig. 20. Diagrammatic representations of four stages in the invasion
of a lake by sapropelic accumulations (which are lightly shaded)
and by the marginal, marshy or aquatic vegetation.　After
Schmitz.

certain lakes in Germany and elsewhere, especially those whose waters are more or less stagnant, and bordered by marshes, or other water-logged lands. These lakes in course of time become silted up, partly by the vigorous growth of surface-living organisms, especially plankton, and partly by the encroachment of the vegetation of the marshy ground bordering the lake. Schmitz (34) has given us an excellent series of diagrammatic views of the stages in the conversion of such a lake into dry land, which is here reproduced (Fig. 20). In the first stage, we see the surface plankton, and the debris of the dead plankton lying at the bottom of the lake basin (shown by light shading). In the second stage, the accumulation on the floor of the lake has increased in thickness, and the marsh vegetation is encroaching on the borders of the lake. In the third stage, both these processes are further advanced, while, in the last stage, the lake has become dry land, and its deposits are of two distinct types, that derived from a plankton accumulation below, and that from a marsh flora above.

In comparison with such a modern accumulation, we may recall the conclusions to which we have been led in recent years, as to the origin and mode of formation of Sapropelic coals, such as Cannels and Bogheads, which we have already discussed (pp. 17, 49). The former are largely made up of micro- and megaspores,

set in a structureless matrix. According to Bertrand
and Renault (2—5), the Bogheads consist chiefly of
aggregations of the thalli of the gelatinous algae, *Pila*
and *Reinschia*, while Jeffrey (23) asserts that they,
like the Cannels, are spore coals pure and simple. It
is obvious that the spores, so abundant in these types
of coals, must have originated from the forests, chiefly
composed of Lycopods, which then existed not very
far from the position in which the coals are now
found. These spores were no doubt accumulated in
water, and very probably were carried a short distance
by streams, in some cases into lakes. If *Pila* and
Reinschia be algae, then they no doubt formed part
of the plankton of the still waters of some lake,
which also received contributions of pollen and
megaspores from the forests of the marshy ground,
by which it was bordered. We are thus, as Bertrand
and others have pointed out, led to the conclusion
that the most probable set of conditions under which
the Sapropelic coals were formed was in the stagnant
waters of a lake.

We have still, however, to account for the brown,
opaque matrix, in which these objects are found
embedded in the Sapropelic coals.

Potonié (28) has shown that on the floors and
surfaces of some lakes to-day there is accumulating
a substance, which he terms *Sapropel*. This is a
gelatinous mass, or slime, which results from the

former plankton, both animal and vegetable, partly decomposed or putrified in quiet or stagnant water. In addition to algae, diatoms, and pollen, small animals such as Crustacea and their excretions, contribute to the mass of the Sapropel, and it is this substance which forms the matrix in which the least destructible organs, such as spores, finally become embedded.

Potonié believes that, in the case of Cannel coals, we are dealing with a fossil Sapropel, largely derived from pure vegetable material, especially algae and spores. He explains the Bogheads on the assumption that the gelatinous thalli of the algae, which it is asserted contribute greatly to the bulk of this coal, have become impregnated with Sapropel, and so are to some extent preserved to-day. Sapropel is regarded as having been largely formed by the decay of the algal plankton, and to this source Potonié also attributes the genesis of petroleum. He terms any rock in which Sapropel forms a considerable part, a Sapropelite, and, since the nature of the plankton varies in different localities and under various circumstances, so the resultant rock also varies, and the final product is not always a fuel. For instance, where the Sapropel consists chiefly of diatoms, the resulting Sapropelite is of a siliceous nature. Where however a carbonaceous rock is formed, Potonié (28) terms it a 'Sapranthrakon' if it is of Palaeozoic, and 'Saprodil' if of Tertiary age.

There is no doubt that this is the best explanation of the origin of the matrix of the Sapropelic coals, which has yet been put forward, and it has met with considerable acceptance. It by no means follows that the ground substance of the Humic coals is of a similar nature, though White (42) has suggested that a completely altered, algal plankton may have contributed to the structureless matrix of all coals.

The particular type of lacustrine accumulation, which we have so far considered, no doubt originated from a growth of vegetation in place. But it must be remembered that lakes are silted up by material drifted from a distance, especially in times of flood (Fig. 1), and there is no reason to suppose that deposits of drifted vegetable material do not sometimes occur, in addition to other types of sediments. Among the older rocks, there is considerable evidence to show that the coals of the basins of St Étienne and Commentry in Central France originated from drifted materials deposited in lakes (10 and 16). The latter coalfield is remarkable for its open-air workings, which offer facilities for the study of the rocks, unrivalled elsewhere. Both these basins have been examined in great detail by MM. Grand'Eury (16) and Fayol (10), especially with reference to the conditions of formation of the seams.

The evidence turns chiefly on the mode of occurrence of numerous trunks of trees in the sandstones, associated with the coal seams. These stems are never found in the coal itself, though they abound in the underclays and roofs of the seams. They are entirely without roots, and are frequently cut off square by the seam above them. Some of the trunks are vertical, but Fayol states that the number of prone stems is a hundred times greater than that of the erect trunks. With the stems are also associated leaves, cones and other fragments of vegetation. One trunk is described as occurring upside down, with its base uppermost. The relation of the sandstones to the trunks and seams is also of interest. The former curve upwards and arch over the trunks, in a way which suggests deposition of drifted matter. The stratification of the roof of the coals is not parallel to the seam. While the coals are more or less horizontal, the beds forming the roof have a considerable dip. Some of the seams, when traced laterally, are found to split, which we have seen also happens in the case of coals in South Staffordshire, and elsewhere in this country (p. 112).

M. Grand'Eury and others have concluded that, in these French coalfields, the evidence is all against the view that the mother substance of the coals originated *in situ*. They regarded it as probable that the whole of the materials were transported and drifted into a large lake basin by river agency. Where the stream entered the lake, its current slackened, and the sediments were deposited in their natural order, according to their specific gravities (Fig. 1). The sands and gravels were first deposited and then the muds, but the comparatively light, though bulky, mass of vegetable material would be carried on into its deeper waters, away from the shores of the lake, finally accumulating on its more or less even floor, as a sheet of nearly pure, plant debris.

It may be that this is a true picture of what has actually happened in these regions in Central France. At the same time it is obvious that the evidence is, for the most part, derived from the associated rocks, and not from the coals themselves. Thus, as we have pointed out on p. 111, the objection may be raised that the history of the coals may not have corresponded with that of the rocks above and below. The sandstones with their included trunks may well be drifted sediments, but it does not follow that the coals themselves may not have been at least in part derived from an aquatic vegetation *in situ*. At any rate we have clear evidence at Commentry of the origin of coals under lacustrine conditions.

CHAPTER VI

THE MODE OF CONVERSION OF THE ORIGINAL
MOTHER SUBSTANCE INTO COAL.
CONCLUSIONS

WE now approach the question, how was the original mother substance actually converted into coal? We will accept as a primary hypothesis the view that the accumulation or deposit, which is now represented by a coal seam, originally consisted, almost entirely, of vegetable material.

One of the earliest answers to this question is that offered by the Peat-to-Anthracite theory, to which we have already devoted some attention (p. 68). We have seen that there appear to be grave objections to this explanation, and we have now to discuss other evidence, which would seem to invalidate it completely.

A more delicate analysis of the problems with which we are here concerned tends to show that *length of time* is the factor, which, more than any other, enters into such considerations. If the Peat-to-Anthracite theory holds, it is presumed that the process was extremely slow, and that long intervals between the different stages were necessary, in order that the gradual changes from one type of fuel to

another might take place. In a modern peat-moss
to-day we see some steps in this supposed sequence
in progress, and the time occupied in passing from
one stage to another is undoubtedly very great.

For some years past, however, evidence has been
gradually accumulating, which tends to show that
Palaeozoic coals, at any rate, were formed rapidly.
We now know that examples of coal conglomerates
and coal breccias do sometimes occur in the Carboniferous series, though they may be somewhat rare
phenomena. Perhaps they are more frequent than
is commonly supposed, for hitherto their significance
has not been generally recognised. Thus little
attention has been paid to them, and they may, in
many cases, have been overlooked. The earliest
recorded instances in this country are from the
South Wales coalfield. Logan (26) in 1840, described
coal pebbles, four inches in diameter, from the Pennant
Grit of this district. They have since been recorded
from the Pennant Grits and the measures above them,
of the Bristol coalfield. The lumps of coals here are
angular, the angles being only slightly rounded, so
that they are of the nature of a breccia, rather
than a conglomerate. Thus they have obviously not
travelled far, and they are regarded as having been
derived from the lower measures beneath the grit.
At any rate coal is far too friable a material ever to
withstand transportation from afar.

Dr Walcot Gibson has informed the author that he has observed similar cases, though on a small scale, in the Upper Carboniferous series of the Midland coalfields. The writer has also recently had an opportunity of examining a breccia, composed of large fragments of coal, set in a sandstone matrix, which was passed through in the Waldershare boring in Kent.

From the French coalfields of Commentry and Champagnac, Fayol and others have recorded well-marked breccias, consisting of fragments of coal of an angular nature, as large as a man's fist. Rolled fragments are also associated.

It is an obvious conclusion from such evidence that coal was not only in existence in Carboniferous times, but that the seams, like the associated beds, in some cases underwent denudation before the close of that period. Thus Palaeozoic coals were undoubtedly formed quickly. In Westphalian times (see p. x) in England, the mother substance of the coal-seams was not only accumulated, but the coal was formed and denuded, all within a comparatively short geological period. The sequence of events recorded in the Stephanian rocks of Central France is also exactly parallel to that which we find has taken place at an earlier stage in Britain.

We have further evidence of rapid formation in the phenomena, sometimes met with in coal workings in this country, which are known to the miners as 'wash-outs' or 'horses.' It is occasionally found that a seam of coal passes laterally into sandstones and shales, or conglomerates, which continue for some distance, beyond which the seam again reappears. These wash-outs are the old Carboniferous river valleys, eroded in the Coal Measure landscape, and now choked with fluviatile debris mixed with fragments of coal. One of the best-known examples occurs in the Forest of Dean in Gloucestershire, others are found in North Wales, and they are especially abundant in South Yorkshire. They afford further evidence of the rapid formation of coal, and its subsequent denudation, within a comparatively short period after the rocks were laid down.

Such evidence as we have discussed, pointing to the rapid formation of coal, is opposed to the truth of the Peat-to-Anthracite theory, which implies that long periods of time have been necessary for even one type of fuel to pass to another. The more modern view, which we may term the Metamorphic theory, is quite independent of lengthy periods of time. It asserts that Palaeozoic coals have resulted from two distinct types of change effected in the mother substance, the first biochemical, and the second metamorphic, and that such fuels are the products of both changes combined.

Biochemical Changes.

It is admitted that the mother substance of coals has accumulated under aqueous conditions. The plant debris, preserved from decay by immersion in water, has nevertheless suffered alteration, under the influence of Bacteria, and possibly also of Fungi. It is imagined that the principal change has been a deoxygenation and dehydrogenation of the material, with a consequent increase in the carbon percentage. Sometimes a marked alteration in the chemical constitution is found, while in other cases comparatively little change appears to have taken place.

In the case of the plant petrifactions occurring in the Lower Coal Measures of Lancashire and Yorkshire (see p. 106), we appear to have some indication of the preliminary changes brought about by such agencies in the mother substance, shortly after its accumulation, but before its conversion into coal. Although the plant remains in the coal-balls are often well preserved, yet, in even the best specimens, portions of the thin-walled tissues, such as the pith, phloem, and inner cortex have disappeared, and a considerable quantity of a brown, structureless substance is present. In other cases, only the resistent woody tissues or the corky elements of the periderm remain, and here the amount of the dark

coloured, dense, amorphous material is sometimes
very noticeable. It is inferred that such changes,
paving the way to the act of metamorphism into coal,
have been brought about by the action of Bacteria in

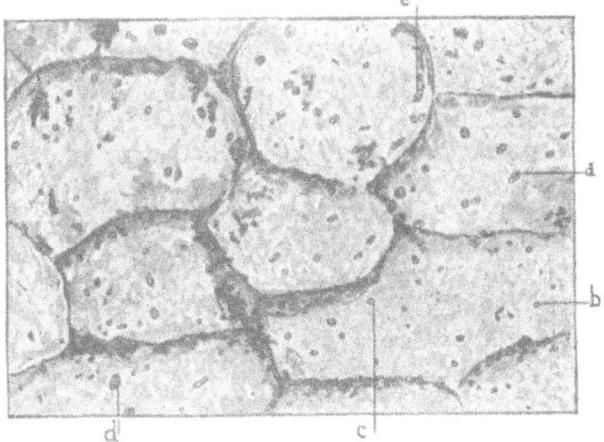

Fig. 21. *Micrococcus zeilleri*, a bacterium in the epidermal cells of
the bark of a *Bothrodendron* from the Toula coal (Upper Carboniferous) of Russia. *a, Micrococci* disposed in line simulating
a *Bacillus*; *b,* Isolated *Micrococcus*; *c, Micrococcus* undergoing
division; *d,* Colonies of *Micrococci*. Magnified 800 times. After
Renault.

Carboniferous times, just as we have seen that Bacteria
play a special part in the history of a modern peat-
moss. We have therefore to inquire how far the

existence of such extremely microscopic organisms as Bacteria in Carboniferous times can be proved.

It would seem that even the most critical must now admit that at least a fair case can be made out, pointing to such a conclusion. The late Bernard Renault devoted much time and attention to examining petrified Carboniferous tissues, and other fossil objects likely to show evidence of these organisms. Some of his results (29—31) appear to be beyond dispute, particularly in the case of some coprolites or fossilised excreta of Permo-Carboniferous fish or reptiles, in which he and Prof. Bertrand have demonstrated the actual Bacillus. The epidermal cells of *Bothrodendron* forming the Toula coal (see p. 42), to which we have already referred, were believed by Renault (30) to show specimens of Micrococci in the petrified state (Fig. 21). Some other cases, described by the same author as being of a like nature, are more doubtful, and it may be that in certain instances he has mistaken specks of mineral origin, perhaps introduced in the cutting of the sections, for the remains of Carboniferous Bacteria

Apart, however, from the identification of actual Bacteria in Palaeozoic rocks, Renault has deduced, from a detailed study of plant petrifactions, well-nigh conclusive evidence to show that the tissues, included in these petrifactions, exhibit all the indications of cells attacked and destroyed by Bacteria, with which one is familiar in recent plants.

There are other reasons for inferring that the accumulated mother substance must have undergone considerable alteration before its actual conversion into coal. In some cases at least this accumulation consisted originally of a great variety of plants, and a mixture of all their detached organs, such as cones and leaves. The first step must obviously have been the conversion of a heterogeneous assemblage of plant debris into a more or less structureless mass. No doubt the completeness of this process varied in different cases, and, in some instances, all trace of the original structure has disappeared, except the cutinised walls of epidermal cells, sporangia and spores, which are almost indestructible. As we have seen, it is extremely probable that this condition was produced by the action of Bacteria, for it appears to be very unlikely that regional metamorphism alone would produce such an effect. It is quite reasonable to suppose that bacterial decay would also cause profound changes in the chemical nature of the mass, though it is not clear at present what influence, if any, regional metamorphism may have had in altering the chemical composition—a much disputed question at the present time.

We have also to account for the origin of the brown or black opaque ground substance of the coal. This matrix permeates the whole mass, and any fragments of tissue still existing in the coal. As we have pointed out above, we seem to detect this substance even in the plant petrifactions, which have only undergone some slight decay, and the obvious conclusion is that it is the final product of the destruction of the tissues by Bacteria. At a more advanced stage the ground substance, consisting of humic, ulmic and other organic acids, becomes more and more abundant, and eventually contributes to the greater portion of the bulk of a Humic coal. Such a product was undoubtedly formed by biochemical changes rather than by regional metamorphism, and must have existed before the act of the complete conversion of the mother substance into coal.

Regional Metamorphism.

According to modern views, the actual conversion into a Humic, Sapropelic or Anthracitic coal of a mass of vegetable material, previously prepared more or less by biochemical changes induced by the influence of Bacteria, is due to regional metamorphism. Thus such coals are essentially metamorphic rocks, and a seam of coal corresponds to a band of metamorphosed limestone. In neither case is the product always the same. A pure limestone may be converted into marble, while an impure calcareous bed may be changed to a garnetiferous rock. It is the same with carbonaceous deposits. A mother substance of one type is metamorphosed to a Humic coal, and another to an Anthracite.

It is known beyond doubt that at the close of Palaeozoic times, the regions now occupied by the English and Welsh coalfields were profoundly affected by a series of great earth movements or crust creeps. As we have already pointed out (p. 111), there is reason to believe that the mother substance of the coals was originally deposited in horizontal or gently inclined sheets of enormous extent, and it is due to the subsequent earth movements that we now find these beds confined to isolated basins. Not only were the rocks as a whole thrown into folds, but the individual beds were frequently faulted, folded or even contorted. If we examine, in the field, Coal Measure deposits which have undergone folding or contortion, we shall gain a good idea of the enormous pressures and lateral thrusts to which they must have been subjected. One could almost imagine that they had been squeezed like putty in the hand of a giant. The effect of the lateral and crushing pressures differs with the type of rock in question. The sandstones and shales associated with the coals are often comparatively little altered structurally, though they bear many external evidences of the pressures which they have undergone. On the other hand, the original mother substance of coal was a comparatively soft, unstable accumulation, and we can readily imagine that more profound internal, as well as external, changes would be produced on such material, as the result of earth movements. The nature of the changes in question appear to have been partly chemical, and partly physical. The devolatilisation of the mother substance, commenced under biochemical processes, may have been completed, and at the same time the rock rendered compact and dense by metamorphism.

Such is the metamorphic theory of the origin of Humic, Sapropelic and Anthracitic coals. There is much to be said in its favour, but at the same time it is necessary to point out that at the present day it is not accepted, in the form here stated, by all geologists. The evidence, on this hand or on that, is very conflicting and puzzling, and, of all the problems bearing on the natural history of coal, we probably know least of the actual means of conversion of the fuel. We are still in the 'dark ages,' so far as this matter is concerned.

One argument in favour of the metamorphic theory, is that in the Cretaceous and Tertiary rocks we rarely find fuels comparable to Humic coals or Anthracites, though Brown coals and Lignites abound. When we examine the geology of these rocks, however, we find, as a rule, no evidence that they have ever

undergone earth movements, similar in kind to those
to which the ancient Palaeozoic rocks have been
subjected. This evidence is not, however, conclusive.
In the Mesozoic rocks true Humic and Anthracitic
coals undoubtedly occur (see p. 35), and while in some
cases the rocks may have been metamorphosed by
earth pressures, it is very doubtful whether this has
always happened. For instance, the Jurassic coal-
bearing rocks of the Yorkshire coast and Sutherland,
(p. 36) which yield fuels approximating to true
Humic coals, certainly show no evidence of having
been disturbed by earth movements.

It may be that effects, similar to crust creep with
its lateral thrusts, may be produced by the crushing
or loading pressure of the weight of an enormous load
of sediments, overlying deeply seated coal-bearing
rocks, but it is doubtful how far this can be
demonstrated. Hilt observed in Westphalia, that
the deeper the seam, the smaller is the percentage
of volatile matter contained in the coal. Hilt's law
applies to Humic coals especially. Were it universally
true we should have good evidence to bring forward
for attributing certain changes to load pressure. But
the known exceptions are almost more numerous than
the rule. In South Wales, the Humic coals may occur
below Anthracites in some portions of the field. In the
Pas-de-Calais coalfield of France, the gas coals lie
below the lean coals, which have a much smaller

percentage of volatile matter. In almost all the larger British coalfields similar exceptions are known.

On the other hand, the evidence derived experimentally by subjecting various carbohydrates to artificial pressures, where it is not conflicting, seems to be opposed to the metamorphic theory, as a pressure effect pure and simple. The experiments of M. Frémy (11) and Prof. Zeiller, who have subjected peat, lignite, wood and cellulose to enormous pressures, agree in indicating that the effects of pressure alone do not convert such substances into Humic coals. On the other hand, M. Spring claims that Peat under a pressure of 6000 atmospheres is converted into a true coal which will give coke after incomplete combustion (9). Where heat in addition to pressure has been employed, M. Frémy (11) has found that hydrocarbons, such as sugar, starch, gums, and ulmic acid derived from vasculose, give a black product analogous to the amorphous matrix of a Humic coal.

There appears to be room for further experimental work in this direction, but, as matters stand at present, if we are right in attributing the actual conversion of the mother substance into coal to the results of earth movements, then heating, as well as pressure effects must have borne a part in effecting the transformation.

On the other hand, some observers assert that

pressure causes little alteration, and are inclined to regard the conversion of the mother substance as due entirely to biochemical changes. If this is so it is curious that, in none of our modern accumulations, where biochemical changes of one sort or another are in progress to-day, do the results resemble the products of the past.

On the whole it would appear that the metamorphic theory, even if we must associate with pressure effects those of heat also, is at least likely to lead us in the right direction, though, at the present stage in our progress we may not have come in sight of the beacon which will show us the channel leading to our goal.

The Origin of Anthracite.

In connection with the metamorphic theory, a discussion of the modern views on the origin of the Anthracite of the South Wales coalfield, may prove instructive. South Wales includes the largest area of productive Coal Measures met with in this country. The coals are of two types, Humic (Bituminous) and Anthracitic. It is the only British coalfield in which coals, which are physically, as well as chemically true Anthracites, are found, though semi-Anthracites occur elsewhere. The Anthracites are developed in the western extremity of the field in

Pembrokeshire, and in the north-western portion of the Coal Measures in Carmarthenshire, Brecknock, and Glamorganshire. The rest of the coalfield furnishes coals of the Humic or semi-Anthracitic types. The passage from a Humic coal to an Anthracite may be traced in a single seam.

Many explanations have been offered to account for the observed facts in South Wales. It might be conjectured that the effects of regional metamorphism have been greater in the Anthracitic area than in the rest of the field. De la Beche, however, as early as 1846, pointed out that there is no evidence to support this view, and that, in other areas, particularly in the Radstock coalfield in Somerset, the rocks are far more contorted than in Pembrokeshire, and yet the coals remain of the Humic type.

Others have been inclined to attribute to the influence of heat, induced by intrusive rocks, the change from the Humic to the Anthracitic state. In the North of England, dykes occur cutting through seams, but at the junctions we always find natural cokes, and not Anthracites, and further, the alteration in the coal is confined to the immediate neighbourhood of the dyke, and is purely local. In the Leicestershire and South Staffordshire coalfields, sheets of igneous intrusions also form coke, where they cut across the seams. In South Wales, the only intrusive rocks are confined to Pembrokeshire, and these are so clearly of Pre-Carboniferous age that they have obviously no bearing on the question of the origin of the Anthracites.

In the recently published memoir on the coals of South Wales, Strahan (40), after an exhaustive survey of the facts bearing on the origin of Anthracite, arrives at the conclusion that the variations in the composition of the coals has been either original, or of very early date. Thus the Anthracites were not derived from Humic coals, as has been often asserted. The mother substances of the two coals differed in the nature of the vegetation of which they consisted, or in the kind of biochemical changes to which they were afterwards subjected. Strahan emphasises the freedom from ash of the Anthracites, as pointing to the existence of a special type of mother substance.

This view will also explain how it is that the two types of fuel in South Wales may occur in one and the same seam. No doubt the mother substance of one portion of the seam differed from that of another. We have already endeavoured to demonstrate that the Humic coals have had a different mode of formation from the Sapropelic, and, as regards the Anthracites also, we appear to have evidence of a third type, which may well have differed in its origin from both the others. In the case of the Anthracites, however, the fuel has become so highly carbonised, and its structure is so homogeneous and opaque, that it is almost impossible to gather, from a study of microscopic sections of this coal, any evidence as to the original nature of its mother substance. It is also a curious fact in this connection that, at present, we have no reason to suspect that the fossil plants, associated with the anthracitic seams in South Wales, differed either as an assemblage, or in their dominant constituents, from the floras found near the Humic coals of the same or other coalfields. If the Anthracites have originated from a type of mother substance peculiar, in this country, to portions of South Wales, we have no idea at present of what this assemblage consisted. Seeing that it is very unlikely that the amount of ash of the original substance would have been altered by any type of biochemical change, we must regard the nature of the vegetation, forming the mother substance, as having determined subsequently the physical and chemical properties of this type of coal.

We have evidence of this fact in the different layers of an ordinary Humic coal (p. 18). The 'mineral charcoal' or 'mother-of-coal' is no doubt formed of different material to the 'bright coal.' The former in all probability consists of altered wood, its charred appearance being due, according to White (42), to the wood having been affected by dry rot, previous to carbonisation, and not to actual charring by heat. The bright coal appears to consist of a mixed assemblage of fragments of leaves, cones and organs other than wood, though no doubt twigs and branches may also occasionally occur.

Conclusions.

In the previous pages we have attempted a brief review of a vast subject, in which an enormous and rapidly increasing literature exists. It is naturally not possible, within the narrow limits of the present volume, to do justice to all that has been said already, or to enter into the details of particular cases which have been cited as having an important bearing on this or that point. We have endeavoured rather to concentrate our attention on the main issues, for it is obvious that the whole set of problems connected with the natural history of coal is extremely complicated. It is equally apparent that the contemplation of one aspect alone may lead to false conclusions.

In the past it has been too often assumed that the evidence which may be obtained in support of one view, is necessarily hostile to another (35), whereas it may eventually appear that it has really no particular bearing one way or the other. It has been thought, and is still often imagined, that, if the Growth-*in-situ* hypothesis is true, the Drift theory is false, or *vice versa*. A beginning has too often been made on the initial assumption that coals have all arisen under the same circumstances, though when the diversity in the chemical and physical properties of the fuels known by this name is recalled, it is difficult to see why this starting-point should be chosen. Fortunately, however, there have been signs, in recent years, that this supposition is being discarded. Prof. Potonié, one of our greatest authorities on the natural history of coal, though a champion of the *in situ* theory, and unwilling to admit that the mother substance of any coal has accumulated from drifted material, is now one of the chief exponents of the view that coals have been formed under very different circumstances and from very diverse materials. To the writer it appears that this is a great step forward, and that once this conclusion is generally accepted, the question of *drifted versus in situ* accumulation or deposition becomes of very secondary importance. If we admit that some coals have originated under terrestrial conditions like modern Peats, while others have been formed in estuarine and lacustrine environments, it is hardly possible, in view of the processes of nature which may be studied in progress to-day, to deny that in some cases coals have been formed *in situ*, while in others the mother substance was transported from a distance to the present site of the seam. There is also very little doubt that some coals have been formed partly from drifted material, and partly from vegetation which grew in place. Further, in some cases at least, the

products which result may be similar, though in one
case the mode of accumulation of the mother substance
may have been terrestrial, and in another estuarine.

We have seen that there is no such thing as a
standard coal, and that in such a consideration we
must take into account the origin of at least six
varieties of carbonaceous rock known by this name.
These substances not only vary in age, from Devonian to Recent, but the rocks with which they are
associated have been formed under very diverse
conditions. In regard also to subsequent alteration,
as the result of earth movements, their histories have
often been very dissimilar. It is therefore very
unlikely that all coals have been formed in the same
way.

A seam of coal represents but one of a sequence
of epochs, which must be studied as a whole if we are
to understand exactly what the circumstances were at
the time of its formation. On the other hand, it does
not follow, as has been so often assumed, that the
conditions under which the seam accumulated were
identical with those of the underclay below, or of the
roof above. We have emphasised the fact the roof
may consist entirely of materials of drifted origin,
while the coal below may have been formed from
material which grew in place. Further the presence
of an underclay (corresponding to a modern estuarine
ooze though not necessarily a Tropical deposit), with

its abundant Stigmarian rhizophores, many of which are *in situ*, is no argument that the coal above has also been derived from vegetable material *in situ*, and not transported.

We have called attention to the evidence of the sorting and sifting action of tides and currents on the accumulating mass of plant debris, as affording a simple explanation of many of the observed facts, and the conflicting conclusions drawn from them. Under such conditions a mixed assemblage of plant debris is not necessarily of transported origin. Again, the evidence of upright trunks in the rocks associated with coal seams is not conclusive that the seams have been formed *in situ*. Some of these trunks in one locality may be in place, while in another they have been transported, and it is difficult in many cases to determine what inference should be drawn from their occurrence.

With regard to the character of the vegetation contributing to coal seams, there is still a great difference of opinion. Grand'Eury (17) has gone so far as to conclude that all coal seams of whatever age were formed from marsh plants. Henslow (22) has recently suggested that upland plants, as well as those of aquatic or swamp habit, have contributed to the bulk of Palaeozoic coal seams, and this view has been upheld here (p. 116). It is to be hoped that the more detailed studies of Carboniferous plants, from the

point of view of their physiological anatomy, now in
progress will throw further light on this question.

We have in recent years reverted to the old view
that the differences in the physical and chemical
characters of coals are mainly due to differences in
the plants which formed the original mother substance.
Lindley and Hutton in their *Fossil Flora* (Vol. ii,
pl. xxvii) in 1833 and Goeppert (13) in 1848 had
already arrived at this conclusion. The modern view
is that the nature of the vegetation composing the
mother substance determines the type of coal, and
that Sapropelic, Humic and Anthracitic coals have
originated from accumulations, quite distinct from
one another, as regards the kind of plant debris in
each case.

We have concluded that the processes, which in
Palaeozoic times, caused the actual conversion of the
mother substance into coal were two-fold, firstly a
biochemical change, secondly a metamorphic act.
Some authorities attribute the greater influence to
the former, and it must certainly be admitted that the
degree and precise nature of the biochemical change,
may have in many cases profoundly influenced the
final product, as much so perhaps as the original
nature of the mother substance. But the evidence we
have that the Palaeozoic types of coal were formed
comparatively quickly, and that the mother substance
did not pass through all the stages postulated by the

Peat-to-Anthracite theory, leads us to attribute a considerable share in the coal building of the Humic, Sapropelic and Anthracitic types to regional, if not thermal, metamorphism.

From such considerations as these, it is obvious that the carbonaceous rocks, which we term coals, were formed in very different ways, under very different conditions, and from different accumulations of plants, belonging to many distinct classes in the vegetable kingdom.   Thus we must study each set of coal-bearing rocks, and each seam of coal on its own merits, if we wish to ascertain the natural history of the coal in question.

BIBLIOGRAPHY

1. Barrois, C. Sur le mode de formation de la Houille du Pas-de-Calais. Ann. Soc. Géol. du Nord, Vol. xxxiii, p. 156. 1904.
2. Bertrand, C. E. Nouvelles remarques sur le Kerosene Shale de la Nouvelle Galles du Sud. Bull. Soc. Hist. Nat. Autun, Vol. ix, p. 193. 1896.
3. —— Notions nouvelles sur la formation des Charbons de Terre. Rev. du Mois, Vol. iii, p. 323. 1907.
4. Bertrand, C. E. and Renault, B. *Pila bibractensis* et le Boghead d'Autun. Bull. Soc. Hist. Nat. Autun, Vol. v, p. 159. 1892
5. —— *Reinschia australis* et Premières remarques sur le Kerosene Shale de la Nouvelle Galles du Sud. Bull. Soc. Hist. Nat. Autun, Vol. vi, p. 321. 1893.
6. Dawson, J. W. On the Coal-Measures of the South Joggins, Nova Scotia. Quart. Journ. Geol. Soc., Vol. x, p. 1. 1854.
7. —— On the Vegetable Structures in Coal. Quart. Journ. Geol. Soc., Vol. xv, p. 626. 1859.
8. de Lapparent, A. L'Origine de la Houille. Rev. Quest. Scientif, Sér. 2, Vol. ii, p. 5. 1892.
9. —— Traité de Géologie. 5th Edition, Paris, pp. 976—990. 1906.
10. Fayol, H. Études sur le terrain houiller de Commentry. Lithologie et stratigraphie. Bull. Soc. Indust. Mines, Sér. 2, Vol. xv, livr. 3-4, p. 1. 1887.

154 BIBLIOGRAPHY

11. Frémy, E. Recherches chimiques sur le formation de la houille. Compt. Rend., Vol. LXXXVIII, p. 1048. 1879.

12. Gibson, W. The Geology of Coal and Coal-mining. London 1908.

13. Gœppert, H. R. Abhandlung als Antwort auf die Preisfrage—Man suche durch genaue Untersuchungen darzuthun, ob die Steinkohlenlager aus Pflanzen entstanden sind, etc. Naturk. Verhandl. Holland. Maatsch. Wetensch. Haarlem, Ser. 2, Vol. IV, p. 1. 1848.

14. Gothan, W. Untersuchungen über die Entstehung der Lias-Steinkohlenflöze bei Fünfkirchen (Pécs, Ungarn). Sitzungsb. K. Preuss. Akad. Wissen. Berlin, Vol. VIII, p. 129. 1910.

15. Grand'Eury, F. C. Mémoire sur la Flore carbonifère du Département de la Loire et du centre de la France. Mém. Prés. Div. Sav. Acad. Sci. France, Sci. Math. et Phys., Vol. XXIV. 1877.

16. —— Mémoire sur la formation de la Houille. Ann. des Mines, Sér. 8, Mém., Vol. I, p. 99. 1882.

17. —— Sur le caractère paludéen des plantes qui ont formé les combustibles fossiles de tout âge. Compt. Rend. Acad. Sci., Vol. CXXXVIII, p. 666. 1904.

18. —— Sur les conditions générales et l'unité de formation des combustibles minéraux de tout âge et de toute espèce. Compt. Rend. Acad. Sci. Vol. CXXXVIII, p. 740. 1904.

19. Green and others. Coal. Its history and uses. By Professors Green, Miall, Thorpe, Rücker and Marshall. Edited by Prof. Thorpe. London, 1878.

20. Gresley, W. S. Notes on the Formation of Coal-Seams, as suggested by evidence collected chiefly in the Leicestershire and South Derbyshire Coal-fields. Quart. Journ. Geol. Soc., Vol. XLIII, p. 671. 1887.

21. Gümbel, C. W. von. Beiträge zur Kenntniss der Texturverhältnisse der Mineralkohlen. Sitzungsber. K. Bayer. Acad. Wiss. München. Math.-Phys. Cl., Vol. XIII, p. 111. 1883.

22. Henslow, G. On the Xerophytic characters of certain Coal-plants, and a Suggested Origin of Coal Beds. Quart. Journ. Geol. Soc., Vol. LXIII, p. 282. 1907.

23. Jeffrey, E. C. On the nature of so-called Algal or Boghead Coals. Rhodora, Vol. XI, p. 61. 1909.

24. Lewis, F. J. The Plant Remains in the Scottish Peat Mosses. Parts I–III. Trans. Roy. Soc. Edinburgh, Vol. XLI, Pt III, p. 699, 1905; Vol. XLV, Pt II, p. 335, 1906; Vol. XLVI, Pt I, p. 33, 1907. For Summary, see Sci. Progr., (New Ser.) Vol. II, No. 6, p. 307, 1907; and Scott. Geogr. Mag., Vol. XXII, p. 241. 1906.

25. Link, H. Über den Ursprung der Steinkohlen und Braun-kohlen nach microskopischen Untersuchungen. Abhandl. Akad. Wissen. Berlin, Vol. for 1838, p. 33. 1839.

26. Logan, W. E. On the Characters of the Beds of Clay immediately below the Coal Seams of South Wales, and on the occurrence of Boulders of Coal in the Pennant Grit of that district. Trans. Geol. Soc., Ser. 2, Vol. VI, Pt II, p. 491. 1842.

27. Lyell, Sir C. Travels in North America, Canada, and Nova Scotia, with Geological observations. 2 Vols., 2nd Edition. London, 1855.

28. Potonié, H. Die Entstehung der Steinkohle und der Kau-stobiolithe überhaupt. 5th Edition. Berlin, 1910.

29. Renault, B. Sur quelques Microorganismes des combustibles fossiles. Bull. Soc. Indust. Miner., Ser. 3, Vol. XIII, p. 865, 1899; Vol. XIV, p. 5. 1900.

30. —— Note sur les cuticules de Tovarkovo. Bull. Soc. Hist. Nat. Autun, Vol. VIII, Proc. Verb., p. 136. 1895.

31. —— Recherches sur les Bactériacées fossiles. Ann. Sci. Nat., Sér. 8, Bot. Vol. II, p. 275. 1896.

32. Renier, A. Observations paléontologiques sur le mode de formation du terrain houiller Belge. Ann. Soc. Geol. Belg., Vol. XXXII, Mém. p. M. 261. 1906.

33. Rigaud, F. La formation de la houille. Rev. Scientif., Sér. 4, Vol. II, p. 385. 1894.

34. Schmitz, G. Formation sur place de la houille. Rev. Quest. Scient., Sér. III, Vol. IX, p. 462. 1906.

35. Seward, A. C. Coal: its structure and formation. Sci. Progress [First Ser.], Vol. II, pp. 355 and 431. 1894.

36. Seyler, C. A. The Chemical Classification of Coal. Proc. S. Wales Inst. Engin., Vol. XXI, p. 483, 1900; Vol. XXII, p. 112, 1900–01.

37. Shaler, N. S. General account of the fresh-water morasses of the United States, with a description of the Dismal Swamp District of Virginia and North Carolina. Tenth Ann. Rep. U.S. Geol. Surv., Pt I. Geol., p. 261, 1890.

38. Solms-Laubach, H. Graf zu. Fossil Botany. Eng. Trans. Oxford, 1891.

39. Stopes, M. C. and Watson, D. M. S. On the present distribution and origin of the calcareous concretions in coal seams, known as 'Coal Balls.' Phil. Trans. Roy. Soc., Ser. B, Vol. 200, p. 167. 1908.

40. Strahan, A. and Pollard, W. The Coals of South Wales, with special reference to the origin and distribution of Anthracite. Mem. Geol. Surv. England and Wales, 1908.

41. Toula, F. Die Steinkohlen, ihre Eigenschaften, Vorkommen, Entstehung und nationalökonomische Bedeutung. Vienna, 1888.

42. White, David. Some Problems of the Formation of Coal. Econ. Geol., Vol. III, p. 292. 1908.

INDEX

For EU product safety concerns, contact us at Calle de José Abascal, 56–1°,
28003 Madrid, Spain or eugpsr@cambridge.org.

www.ingramcontent.com/pod-product-compliance
Ingram Content Group UK Ltd.
Pitfield, Milton Keynes, MK11 3LW, UK
UKHW010851090126
466816UK00011B/146